TANK CAR Color Guide

Volume 1: Cars with Full Center Sills

Wyandotte tank cars; April 1960, Wyandotte, Michigan *(Emery Gulash, Morning Sun Books collection)*

by **James Kinkaid**

Copyright © 2010
Morning Sun Books, Inc.

All rights reserved. This book may not be reproduced in part or in whole without written permission from the publisher, except in the case of brief quotations or reproductions of the cover for the purposes of review.

Library of Congress
Catalog Card No. 2010928147

First Printing
ISBN 1-58248-311-6

Published by
Morning Sun Books, Inc.
9 Pheasant Lane
Scotch Plains, NJ 07076
Printed in Korea

Robert J. Yanosey, President
To access our full library *In Color* visit us at
www.morningsunbooks.com

Dedication

I would like to dedicate this book to Craig Bossler. Craig was one of my early contacts many years back when I first became interested in freight cars and his immediate, and much appreciated (to this day), gift of assistance with helping me to begin my library started me on this path which I still travel. It is doubtful that I would have proceeded down it had it not been for his generosity, which extends to this volume.

Thanks, Craig!

Acknowledgements

I would very much like to acknowledge the assistance of Eric Neubauer. Without his help and assistance, the book would lack a great deal and I am truly happy for his help. I've always thought that tank cars are the most interesting of freight cars and I am thrilled that Bob Yanosey asked me to write these two books. As always, my great thanks to Bob for his help, advice and patience with me and most of all for taking the chance on these two volumes.

As with all Morning Sun Color Guides, this volume would not be possible without the help of many contributors. This is even more so since these two books rely heavily on individual collections. So, I offer my heartfelt thanks to: Gib Allbach, Calvin T. Banse, Jerry Bosanek, Craig Bossler, Emery Gulash, Matt Herson, David Hickcox, Pat Holden, Fred Kite, Norman E. Kohl, Al Lanier, John Luckfield, Owen Leander, John McCown, Dave McKay, Dave Nelson, Don Reck, Jim Rogers, William Rosenberg, Lou Schmitz, Jim Thorington, Bernie Wooller, Paul C. Winters, Robert Yanosey, Lee Yoder and Chuck Yungkurth of Rail Data Service.

Table of Contents

Single Compartment Cars without Platforms	4-19
Non-Insulated Single Compartment Cars	20-44
Insulated Single Compartment Cars	45-90
Small Cars	91-99
6-Axle Cars	100
Multi-Compartment Cars	101-118
AAR Class TMU	119-121
AAR Class TW	122
AAR Class XT	122-127
AAR Class TVI	128

TANK CAR Color Guide
Volume 1: Cars with Full Center Sills

Introduction

This is Volume One of a two volume set and as such, the volumes are meant to be utilized together. It deals with tank cars equipped with full underframe center sills, regardless of car design. Cars without full center sills are presented in Volume Two.

These two books are different than most *Morning Sun Color Guide to Freight and Passenger Car* books in two important ways.

The first deals with the subject matter. In the color guides devoted to a particular railroad the images presented are defined by what the road operated. We can't do this here. The canvass is too big and varied: there are just too many operators, lessors, lessees, reporting marks (which often changed) and too many small production runs (often only a single car), sometimes by builders with limited output. So, here we've chosen to follow where some excellent photographic collections lead us. And since these are all-color books, we're unabashedly focused on colorful and interesting cars (though as will be seen, there are certainly plenty of very interesting plainly painted black or white cars around). The Frae only restrictions here are that the cars must have been in private owner service (meaning the reporting marks end in "X").

The second difference deals with the research. Generally speaking, just about every railroad has its fan base and often there are resources available to help sort out backgrounds on cars. Not so with tank cars: there is no central repository of information. Excepting a partial American Car and Foundry lot list (which lacks some important information itself) there is little "hard" history for this genre. Thus, for the captions herein, I've had to rely on scattered production information, *Railway Age* order information, field reporting and sometimes educated guesswork. I was able to utilize a fairly complete set of *Official Railway Equipment Registers*, but when dealing with some of the major lessors this information is very incomplete and series limits I've given must be used with a certain degree of skepticism. Another caveat: when discussing the major lessors, just because one car in a given series might be under lease to a particular company do not assume that all are. GATX and UTLX tank car manuals and gauging books were also consulted. Yet in several instances, so little is known that I just have to let the picture tell the story.

About the Project

The main question was how to present the material. Some seemingly obvious choices would be by builder, date, car specification, commodity, industry, etc. Much of this data is not available so I decided to sort out the cars from a modeler's perspective, i.e., by major construction features. Within each section cars are presented in strictly alpha-numerical order. This means that ACF's Shippers Car Line cars are reported several times (since they undertook a program to restencil tank cars from SHPX to ACFX over the years). In other instances you'll find cars painted for lessee's such as Diamond Shamrock scattered around within the books. Close inspection will show many car and paint scheme differences, which makes their inclusion worthwhile. Cars may or may not have heater pipes or coils. If this fact is known, it will be noted. Although the existence of coils doesn't normally change the visual appearance of the car, this can be an important detail for those interested.

One note here: all tank cars have some sort of specification markings and these have changed over the years, so I've given what the car had stenciled or reported at the time of its photo, if known.

About Tank Cars

The modern tank car has been around in this form since the late 1800's. While early efforts were flat cars with barrels attached to the decks, it wasn't long before the advantages of the horizontally laid cylinder became apparent and most future tank cars were designed that way. For years, the tank car was a rather simple riveted affair primarily used to move crude oil and gasoline. By the late 1920's, concurrent with the ICC taking over jurisdiction for hazardous material transport, tank cars began to be much more specialized.

Tank cars, like refrigerator cars, are treated differently by the railroads who do not normally own revenue tank cars (though there are exceptions) since they are so specialized and usually in dedicated service. Thus individual owners must operate their own cars, which has led to the rise of a number of leasing companies that supplied cars for those not wishing to buy cars outright.

The huge increase in equipment designed for specialized services and the constantly changing leasing situation has led directly to these two books. The variety is enormous and seemingly unending.

Tank cars have a dual nature about them in that they are given classes by the AAR and also specifications issued by a regulatory body.

The AAR tank car classes, primarily for equipment register use, are of a general nature describing the overall function of the car. These are: TA (acid service), TL (lined cars), TM (general service), TMU (multi-unit tank cars), TP for pressure cars (generally meant to apply to cars with a tank working pressure of over 100 psi), TR (rubber lined), TV (cryogen cars that typically used an inner and outer tank separated by a vacuum), TW (wooden tanks) and XT (box cars with internal tankage). The character "I" could be added to signify that the car was insulated.

However, the AAR coding is much too simplified for precision use. Each tank car also has a specification, which is a specific reference that governs the mechanical features of the pressure vessel and its appurtenances, tank manufacture, fittings and structural characteristics and from that, what commodities can be carried. For non-hazardous commodities, these are issued by the AAR. For tank cars utilized in hazardous work, the ICC (and after April 1, 1967, the U.S. DOT) authorized the specifications. Between January 1, 1903, when the Master Car Builders' Association first published standards for tank car construction and repair, up to January 1971 there have been 127 different tank car specifications issued.

Single Compartment Cars without Platforms

AESX 201

▶ The A. E. Staley Company of Decatur, Illinois has operated tank cars under this reporting mark since at least 1913. A 40-ton capacity car, it was one of 107 acquired in the mid-1940's and numbered 183-260, which were added to the previously existing 20-182 set. Its history is unknown, but was still in service when photographed on May 6, 1962 at Columbus, Ohio.
(Paul C. Winters)

AESX 719
ARAIII
▲ Marked with a March 1918 manufacture date the car is mismarked as "AAR" specification III. A 50-ton capacity car, it was one of about 66 cars first mentioned in July 1960 and marked AESX 676-740, which were added to the previous 50-ton cars in AESX 15-675 (40-ton cars were also rostered). It is thought that these cars were acquired used from North American, although the exact lineage is unknown. Photographed in August 1967 it was built by the Standard Car Construction Company. *(William Rosenberg)*

AESX 790
ICC103
◀ Coupled to 201 above, car 790 was a car placed in service in October 1961 as part of the AESX 741-840 group. Like the other two cars illustrated here, it was a nominal 8,000 gallon car. Eventually Staley combined all of these cars into one contiguous series, making it a mixed group, with 41 cars remaining until January 1983 after which they were purged from the register. Note that all three of these cars were non-insulated and had heater coils fitted.
(Paul C. Winters)

AGAX 236
ICC105A300W
▲ An insulated car operated by the Anchor Gasoline Corporation sits at Council Bluffs, Iowa on May 14, 1961. In liquid nitrogen fertilizer service it was one of 25 fabricated by American Car and Foundry in February 1954. Numbered 225-249 and from lot 02-4200A, the cars lasted until late 1966 after which Anchor Gasoline's reporting marks disappeared from the equipment register. *(Lou Schmitz)*

CCBX 2307
ICC103
▶ Modelers would have a challenging job reproducing this Union Carbide car, spotted in May 1968. Markings indicate that it was constructed by General American Transportation Company. While the manufacture date is unreadable, its series (2000-2400) was first listed in the October 1963 *Official Railway Equipment Register*. It was non-insulated and fitted with heater coils.
(Paul C. Winters)

CPVX 111
ICC103W
◀ Sitting out of service in Baldwin City, Kansas on June 5, 1993 it has markings indicating that American Car and Foundry constructed it in April 1920. However, because the car was not listed until circa 1940, it was apparently acquired used. The car was retired in the summer of 1970. *(Jim Kinkaid)*

DODX 12099 & 16414
DOT103W

▲▼ Two ex-United States Army cars. When the various Department of Defense cars began to be consolidated under "DODX" reporting marks in October 1965, they were restenciled. In the upper illustration is a car constructed by American Car and Foundry. It was originally part of the United States Army series USAX 11900-12099. That set of cars was assembled at Milton, Pennsylvania in September and October 1949 under lot TC-3410 as 10,177 gallon capacity cars. It was photographed at Memphis, Tennessee on September 27, 1983. At middle is a car originally part of a 25-car series numbered USAX 16414-16438. It was constructed in May 1952 by General American Transportation and is illustrated while at Atlanta, Georgia on April 12, 1988. Note the all-welded tanks (thus the suffix "W" in the specification) on these non-insulated cars.

*(top, Al Lanier, Jim Kinkaid collection;
middle, Jim Kinkaid collection)*

GATX 11
DOT103

▶ Photographed in August 1973 it has markings indicating a November 1938 build date. While uninsulated, it does have heater coils; the giveaway are the four sets of three rivets in triangular patterns which are the interior heater coil anchors. The car has a 4,051 gallon capacity.

*(Rail Data Service,
Jim Kinkaid collection)*

GATX 9487
ARAIII
▲The United States Railway Equipment Company photographed this car after performing minor repairs at Blue Island, Illinois in July 1963. The car was repaired there since the lessee, Clark Oil Refinery (originally Clark Oil and Refining, incorporated in 1932 and named for founder Emory Clark) had its refining plant there. It was constructed by American Car and Foundry in March 1920 and was later shown as part of GATX 9480-9499. It was fitted with heater coils and non-insulated.
(USRE photo, Jim Kinkaid collection)

GATX 32678
ICC103W
◀ This insulated car apparently has its original paint from the General American factory at Sharon, Pennsylvania, where it was built in August 1951. First listed in October 1951 as part of the 60-car GATX set 32606-32725 it sits at Council Bluffs, Iowa on October 16, 1955. Victor Chemical Works, by the way, hails from Chicago with origins in 1932 and was later part of Stauffer Chemical Company. *(Lou Schmitz)*

GATX 33673
ICC103W
◀ An insulated car constructed by General American in June 1949, it began to be listed in October 1959 as part of the 48-car GATX 33652-33699 series. Found at McCook, Illinois on April 17, 1965 its lessee was the Celanese Corporation of America, a large producer of acetyl products.
(Owen Leander, Morning Sun Books collection)

GATX 36634
ICC103
▲ An insulated car with coils (notice the fittings sticking out from the left hand tank head), this car had been constructed by General American in August 1942. Photographed in February 1973 it was part of the seven car GATX series 36634-36640. Continental Turpentine & Rosin was incorporated in 1957 and made these products through 1984. *(Dave Nelson, David Hickcox collection)*

GATX 61234
DOT103AW
▲ That unusual dome and corrosive placard board certainly suggest acid service on this gem spotted at Tullahoma, Tennessee on June 1, 1974. Built by General American in July 1951 it was one of 100 in the series 61200-61299. The "AW" within the specification denoted an all-welded tank and that bottom outlet and safety valves were prohibited (although safety vents were required). *(Lee Yoder)*

GATX 76723
ICC103
▶ Although the build date cannot be determined, it is known that the car was first listed in 1952 as part of the insulated and lined GATX series 76710-76749. Fitted with heater coils it was put on film in August 1971. American Maize Products (the company name can be traced back to 1908) shipped many products derived from corn, with corn syrup being a major commodity.

(William Rosenberg)

GATX 95245
ICC105A300W
▲ Warren Petroleum, of Tulsa, Oklahoma (founded in 1922 by William Warren) hauled liquefied petroleum gas in this car. Photographed in April 1963 the car's first specific listing was that January, when it was recorded as part of the GATX series 95236-95249. The car was certainly in existence as far back as May 1948 since that date can be seen. These cars may have been renumbered or acquired used. Its original manufacture date and number is unknown. *(Paul C. Winters)*

GRYX 359
ICC103
◀ The John H. Grace Company owned this car; its lessee, the Franklin Oil Company, was founded circa 1902. It was first listed in the equipment register in early 1926 as one of 12 in the set 300-399, apparently constructed by General American Transportation. An uninsulated car, notice that the heater pipes exit the tank head of this 8,000 gallon car. It was at Cleveland, Ohio in November 1974.
(John McCown photo, Dave McKay, Morning Sun Books collection)

HJMX 101
▶ The Sterling Fuels Company of London, Ontario operated these cars. The car illustrated in this photo, taken in November 1959, was part of the HJMX series 101-103. These cars first appeared in October 1952, which was also when the company began to be listed in the *Official Railway Equipment Register*. It has to have been acquired used since a January 1916 build date is stenciled on the tank.
(Emery Gulash, Morning Sun Books collection)

HJMX 6024

◀ Another Sterling Fuels car found in November 1959. Like car 101 on the prior page, it was an uninsulated car equipped with heater coils. The car was part of the 6019-6028 set added to the equipment register in July 1954. The reporting marks, by the way, stood for H.J. McManus, the owner of Sterling Fuels.

(Emery Gulash, Morning Sun Books collection)

KTX 167
ICC103W

▼ The reporting marks KTX were owned by the Keith Railway Equipment Company. Car 167 was first listed in January 1954 as part of the 100-199 series. The company's reporting marks were transferred to Shippers Car Line in April 1963, where the cars continued to be listed. The 8,000 gallon car was photographed at Birmingham, Alabama in May 1958. It is an uninsulated car fitted with heater pipes. *(Jim Thorington)*

MOBX 8805
ARA III

▶ The Socony Mobil Oil Company (known by this name since at least 1955) registered the reporting marks MOBX in April 1960. This car, manufactured in December 1917, was part of the MOBX series 8805-8811, which were renumbered from another set of cars still unknown. Renamed as simply Mobil Oil in 1966, its owner still had this car in revenue service when found in July 1968!
(Paul C. Winters)

NATX 5350 & 5390
ICC103

▲▼ Mid-1929 saw the manufacture of the North American owned NATX 5300-5499 cars by the Pressed Steel Company. Note that the equipment register showed that there were multi-compartment cars mixed into this group, so there were probably cars of varying designs within the series. Both of these cars happen to have been constructed in May 1929. At top is a photo taken in September 1973 at Huntsville, Alabama on the Southern Railway. And at middle is another caught on film in February 1974 while under lease to Hunt-Wessen Foods, Incorporated. *(both, Bernie Wooller)*

NATX 5957
ICC103

◀ August through October 1929 saw the construction of another 500 cars by the Pressed Steel Car Company for North American. The cars were numbered 5500-5999. Our example was photographed in February 1974 and has markings indicating that it was fabricated in September 1929. All three of the cars on this page were non-insulated and fitted with heater coils.

(Bernie Wooller)

NATX 18156
ICC103W
▼ Here's a photo taken in July 1968. Although most likely built in mid-1957 it was first listed under a separate listing in January 1958 as part of the NATX 18152-18170 set of cars. An uninsulated car it has steam coils for its commodity. Worthy of note are the handrail supports at each corner of the car. *(Paul C. Winters)*

NATX 18500
▶ This car was insulated for the same product and lessee (the J.C. Hubinger Company was a large producer of corn starch and syrup). And it too had heater coils. It was photographed in August 1967 and was part of a set of cars first described in the October 1957 register issue as NATX 18500-18550. The series, however, only contained 10 cars. *(William Rosenberg)*

ROX 152
ARA III
◀ The reporting mark ROX was originally owned by the Richfield Oil Company (in 1966 renamed as Atlantic Richfield) when this car was first listed in January 1939. It was most likely renumbered from the ROX 10000-10060 set of tank cars at that time. Spotted at St. Louis on February 17, 1963 its markings show that it was constructed in February 1925, apparently by General American. At the time of the photo it likely belonged to the set 147-153.

(Paul C. Winters)

SBIX 2413
AAR203W
▲ Standard Brands Incorporated picked up 24 of these cars in either late 1959 or early 1960. The non-insulated cars, which were first listed in the October 1959 equipment register, were numbered 2400-2423. This 8,081 gallon car was at South Modena, Pennsylvania when found on December 29, 1974, likely there for scrapping. Note the two stenciled build dates: the more likely is the February 1960 one. Cars built to this specification were truly non-pressure cars and hydrostatic testing of the tanks was not required. *(Craig Bossler)*

SDRX 15849
ARAIII
◀ The Sinclair Refining Company operated cars under this reporting mark. Assembled by the Pennsylvania Tank Car Company in January 1918, by early 1920 the 8,000 gallon car was part of the SDRX set 15800-15899. The small tag on the route board said that it was loaded with gasoline when photographed in August 1945.

(Emery Gulash, Morning Sun Books collection)

SHPX 4133
ICC103
▼ American Car and Foundry's Shippers Car Line provided this car, found at Hicksville, New York in February 1959. This uninsulated car only carried 4,056 gallons. It was one of 10 in the series 4133-4142 manufactured in August 1937. The lessee had just recently (in 1957) been renamed Union Carbide (from the Union Carbide and Carbon Company). Extra rivets along the lower tank shell indicate internal heater coils.

(Norman E. Kohl, Jim Kinkaid collection)

SHPX 6204
ICC103W
▲ This particular one was at an unmarked location when it was photographed in April 1969. ACF built the non-insulated car in November 1947, possibly under lot 3219, and placed it into the 6200-6204 set. Lessor FMC, previously known as the Food Machinery and Chemical Corporation and renamed in 1948, used the car at their Nitro, West Virginia plant where pesticides were manufactured. *(Paul C. Winters)*

SHPX 8768 & 9503
ICC103
▼ Two run of the mill cars found in St. Louis in September 1967. ACF constructed car 8768 in November 1928 as part of the group 8754-8768. The 15 cars were of 8,138 gallon capacity and built under lot 767. It is hard to make the information out accurately, but companion 9503 would seem to have been built in February 1936. *(Paul C. Winters)*

SHPX 21653
ARAIII
▶ Here is a car sitting on the Southern Railway in May 1967, probably at Huntsville, Alabama. Although markings indicate that it was supplied by General American in October 1919, it was not until circa late 1940 that the equipment register began to list it separate as part of the 21555-22193 series. Uninsulated it is equipped with heater pipes inside.
(Bernie Wooller)

TWOX 1083 & 2045
ARAIII

◀▼ Assembled in March 1929, at top Tidewater Oil Company 1083 is on the move in September 1967. In July 1930 predecessor Tide Water Oil Sales acquired this car as part of the 1000-1403 series of 10,000 gallon cars. Below that picture is a car found in December 1968. Although furnished from the builder in November 1925 the 2001-2199 group that it belonged to was not listed until the mid-1930's.

(top, William Rosenberg; middle, Paul C. Winters)

USAX 11083
ICC103A

▲ Supplied in December 1942 for ammonium nitrate service the 10,000 gallon car was photographed in February 1968. As constructed it was marked USOX and part of the 11057-11156 series. It was a non-insulated car. The cars were later restenciled either with DODX or USAX reporting marks. *(Paul C. Winters)*

USAX 11735
ICC103W
▲ Built in May 1949 by American Car and Foundry it was photographed in April 1961. A 10,000 gallon capacity uninsulated car, it was supplied under lot 3401 and part of the 11690-11899 set. This set of cars was fabricated under lot TC-3417 and as the "W" specification suffix indicates, had all-welded tanks. *(Emery Gulash, Morning Sun Books collection)*

UTLX 14269
ARAII
▲ Union Tank Car provides the oldest car in either volume; it was constructed in June 1915. Although the tank's markings indicate that it was supplied by the Pennsylvania Railroad, this is quite unlikely. Union Tank Car never bothered to break down their cars into separate groups in the equipment register but by January 1920 it was just one of over 20,000 cars of similar size. In acid service it was found still in revenue service at Huntsville, Alabama in June 1965. *(Bernie Wooller)*

UTLX 17635
ICC103
◀ A photo taken in September 1973 of a car constructed in September 1937. This uninsulated car has its internal heater pipes sticking out of the tank head.

(Bernie Wooller)

UTLX 25012
ARAIII
▶ This insulated car was assembled in October 1923 and caught on film in May 1968. There is evidence that it has heater pipes installed.
(Paul C. Winters)

UTLX 48935
ARAIII
▼ Here's an unusual non-insulated car spotted in March 1968. What makes it unusual is that the tank states that it was manufactured in December 1918 yet markings on the underframe say that it was built in December 1956. This car has probably had the tank transferred to a new underframe to let it continue in service due to previous underframe age or damage.
(Bernie Wooller)

UTLX 49186
ICC103W
▲ An all-welded uninsulated car from American Car and Foundry, it was one of 500 constructed under lot 02-3555A and numbered as UTLX 49000-49499. This 12,000 gallon car sits for its photo in March 1966 and was built in September 1951. *(Paul C. Winters)*

UTLX 69736
▲ This insulated car would appear to have been constructed in June 1956 with a tank supplied by Graver Tank and Manufacturing. It was one of about 150 added to the series 68100-69999 in that time frame. The car was spotted in Columbus, Ohio in April 1963 on the C&O Railroad while under lease to Hercules Powder Company, a large explosives manufacturer. *(Paul C. Winters)*

UTLX 85704
ARAIII
▲ Obviously a very unusual car! Carrying transformer oil, it was found on September 8, 1962 in Columbus, Ohio. The photographer's notes indicate that it was consigned to the Columbus & Southern Ohio Electric Company.
(Paul C. Winters)

UTLX 85925
ARAIII
▶ Another car with the unusual fully enclosed housing and somewhat a mystery. It is marked as a 80,000 lb. capacity car, yet when its photo was taken in December 1962, the register showed no cars of this rating. It was built in July 1921 and as can be seen, it also carries transformer oil.
(William Rosenberg)

WCHX 1701
ARAIII
▼ Mismarked "AAR III" the 8,000 gallon car was constructed in late 1926 by American Car and Foundry. However, the Walter Haffner Company's first listing was not until April 1944 meaning that the car was acquired used. Listed as part of the 1701-1754 set it was photographed in January 1962. Lessee Mobile Rosin Oil Company (formed in 1924) supplied rosin, primarily for the automobile rubber industry. *(Paul C. Winters)*

WRNX 14132 & 14415
ICC103
▼▼ In 1953 the Gulf Oil Company acquired the Warren Petroleum Company. Part of the deal was that Gulf also acquired Warren's WRNX reporting marks. So, here and on the next page are some of the resulting cars which were all uninsulated. At middle is one found at Akron, Ohio on March 24, 1967. Its markings show that it was constructed in October 1930 by the Petroleum Iron Works Company. And in the lower picture is a car fabricated by General American in October 1930. It was filmed in January 1969. Note the poling pockets on the sides at the bolsters. All of these cars were first listed in January 1961 as part of a large group numbered 13158-14478. Obviously cars from different sources comprised this set. At present the source for these cars remains unknown.

(middle, Dave McKay, Morning Sun Books collection; bottom, Paul C. Winters)

WRNX 14510
ICC103
▲ A nice overhead shot of a car found in South Modena, Pennsylvania on New Year's Day 1972. Since the car was here, it was likely awaiting scrapping. It was part of the series described on the previous page. Note that this car has internal heater coils. *(Craig Bossler)*

WRNX 15240
ARAIII
▶ A 10,000 gallon car found in January 1969. It was part of the 15000-15291 series of coiled cars supplied by the Standard Steel Car Company in June 1923. The cars appear to have been restenciled from GRCX 5000-5199. Like the rest of these WRNX cars this group was not listed until January 1961. Notice that it is marked as "AAR SPEC 3."
(William Rosenberg)

WUTX 4522
▶ As alluded to in the introduction, sometimes the image has to tell the whole story. This is such a case: there is no information on this car at all: no date, no location and the car is not listed in the *Official Railway Equipment Register*. With KD brake gear and having been last reweighed in May 1944, it was probably photographed around that time.
(Emery Gulash, Morning Sun Books collection)

Non-Insulated Single Compartment Cars

ACDX 26519, 26885 & 27966
ICC103AW, ICC103B & ICC103AW

▲▼▼ First Union Properties was first listed in April 1963 and was a holding company that folded in cars from Allied Chemical's various divisions. At top is one photographed in April 1967. It was from the set 26385-26734 which had been renumbered from GCX 6385-6734. That group can be traced back to at least January 1939, and perhaps even earlier. Note that this 6,500 gallon acid service car is equipped with National B-1 trucks. In the middle slot is a photo taken on June 27, 1971 at the Spruce Street freight station in Reading, Pennsylvania. Also riding on National type B-1 trucks its specification means that it is rubber lined. Its markings show that the tank was constructed by General American Transportation Company in October 1928 while the underframe was built in October 1938. Part of the 74-car ACDX series 26805-26974, this group was renumbered from GCX 6805-6974. And below is a car put on film in May 1967. It was fabricated by American Car and Foundry in December 1960, but apparently has a May 1960 frame date, and is also in acid service. The car was one of two 9,900 gallon capacity cars added in April 1961 (though the series limits increased by 58 cars!) to make up the GCX 7641-7985 set (until then the series had been 7641-7927). This group was then renumbered ACDX 27641-27985. *(top, Jim Kinkaid collection; middle, Craig Bossler; bottom, Paul C. Winters)*

ACFX 7295
DOT103CW

◀ Restenciled from its original SHPX reporting marks the car sits at Reading, Pennsylvania on March 14, 1982. In nitric acid service, it was fabricated in March 1948 and was placed into the SHPX series 7293-7299. American Car and Foundry supplied these seven cars under lot 3223. Part of this car's specification involves the use of a 2% expansion dome. *(Craig Bossler)*

ACFX 24117
ICC103W

▲ Another rather plain car photographed in April 1978. It was built in October 1958 by American Car and Foundry under lot 71-5229 as part of the SHPX series 24067-24166 and was assigned to Hunt-Wessen Foods. Its steam coil outlets may be seen peeking out from the center sill just to the left of the nearest truck. *(Bernie Wooller)*

ACFX 24246
ICC103W

▼ Here's a car found in August 1971. It was manufactured by ACF in January 1959, possibly as part of the set SHPX 24167-24248. Hidden behind the handrail is the lessee: the Houston Chemical Company. Note the dual placard boards along the side of the car. Of interest is the blue coupler! *(Paul C. Winters)*

ACFX 81533
DOT111A100W1

▶ Portland, Oregon was where this photo was taken on October 7, 1972. It was assembled by ACF in September 1964 and would seem to be one of 14 numbered 81547-81560. The lessor of this 20,883 gallon capacity car was a large independent refiner that supplied many petroleum products.

(Jim Kinkaid collection)

AHMX 110
ICC103BW

▲ A photo taken at an unknown location in July 1973. It was constructed by American Car and Foundry in May 1959 and part of the series 102-112. The car was in muriatic acid service, for which it was rubber lined. Compare this car to ACFX 24117 opposite.

(Rail Data Service, Jim Kinkaid collection)

AUPX 102
ICC103AW

▼ The Austin Powder Company (a manufacturer of industrial explosives who has been in business since 1833, just like the logo says!) was operating this car in October 1968. General American constructed it in July 1958 as a single car order. The 8,087 gallon car was in acid service.

(Paul C. Winters)

BECX 81
AAR201A35W
▲ Another single car order from General American, assembled in August 1948. The original owner was the Buffalo Electro-Chemical Company, thus the reporting mark. The car was photographed in November 1966. *(William Rosenberg)*

BECX 897
ICC103A-ALW
▶ A name change resulted in the company being called Becco Peroxide. Here's another of their cars which had been assembled by General American in September 1955 at Sharon, Pennsylvania and was photographed at the same place and date as BECX 81 above. This car used a welded aluminum tank and was part of the three car set BECX 897-899. Rubber lined it was an 8,000 gallon car. *(William Rosenberg)*

CCIX 3005
ICC103BW
◀ The CCIX reporting mark stands for the Consolidated Chemical Industries Division of the Stauffer Chemical Company. A rubber lined acid service car found in Muskogee, Oklahoma in July 1988 it was assembled in December 1958 by American Car and Foundry as part of the CCIX series 3000-3006. *(Jim Kinkaid)*

DODX 6094
DOT103W
▲ We are trackside at San Luis Obispo, California on September 29, 1984. Originally marked USNX for United States Navy aviation fuel use it was built in August 1949 by General American. The original series was 6000-6199 and they were 10,000 gallon capacity cars.
(Pat Holden, Jim Kinkaid collection)

DUPX 7849
▲ E.I. DuPont de Nemours and Company owns these cars. One was caught at Reading, Pennsylvania on July 23, 1990. The car had just been constructed that March and was one of 15 cars in the series 7835-7849. *(Craig Bossler)*

DUPX 14729
ICC111A60W7
◀ Another DuPont car at Reading, Pennsylvania but on May 5, 1989. It was assembled in June 1985 by Union Tank Car who fabricated the car as part of the set 14725-14731, all for chlorosulfonic acid service.
(Craig Bossler)

DUPX 17002
ICC103AALW
▼ General American is thought to have constructed this car for hydrogen peroxide service in July 1964. As such, it was part of the group 17001-17006. When caught on film in August 1970 it was at Birmingham, Alabama. *(Jim Thorington)*

DUPX 17063
DOT111A60ALW2
▲ A car found at Reading, Pennsylvania on May 18, 1991. It was part of the series 17059-17078 and assembled in December 1984 by builder Riley-Beaird in Shreveport, Louisiana.
(Craig Bossler)

EORX 18554
▶ Here is a Cities Service Pipeline Company car found at Tullahoma, Tennessee in July 1967. It was from the series 18500-18599 which first showed up in April 1960.
(Fred Kite, Lee Yoder collection)

FMLX 1791
ICC103A-ALW
▲ Constructed by General American in July 1964 it sits at Birmingham, Alabama in September 1964. Part of the 1790-1792 set, these rubber lined cars were of 17,500 gallon capacity. *(Jim Thorington)*

FMLX 15023
DOT111A100W1
▼ A car caught on film in September 1972, one of 25 in the series 15000-15024. These 15,000 gallon cars first appeared in January 1964, which also happens to be when the reporting mark was also first listed for owner FMC. *(William Rosenberg)*

GATX 6990
ICC109A200ALW
◀ Manufactured by General American in October 1965 it was at North Little Rock, Arkansas when photographed on December 7, 1985. The car was part of the 10-car set 6987-6996. Bottom outlets were prohibited on cars with this specification.
(Al Lanier, Jim Kinkaid collection)

GATX 10966
ICC103W
▲ A car photographed in February 1964. General American built it in June 1949. *(Paul C. Winters)*

GATX 19750
DOT111A60ALW2
▲ Here is a photo taken in Tampa, Florida on March 11, 1982. The rubber lined car was assembled by General American in October 1980 and out of the GATX 19735-19750 series. The 16-car set was fabricated in July through October 1980. Interox was a maker of hydrogen peroxide and based in Houston, Texas.
*(Emery Gulash,
Morning Sun Books collection)*

GATX 20887
ICC103W
▶ Vulcan Materials Company's chemical division often used these cars to ship sodium hydroxide, otherwise known as caustic soda. This one was found in January 1969.
(William Rosenberg)

GATX 24844
ICC103W
▶ A car photographed in August 1967 that was manufactured by General American in December 1947. Based on the equipment register it would appear to be part of the 55-car set GATX 24800-24854. *(William Rosenberg)*

GATX 26809
ICC111A100W2
▼ In sulfuric acid service, it was assembled by General American in April 1948 and photographed in October 1968. It is hard to say what series it first belonged to, but in January 1965 it began to be listed separately. Possibly renumbered, by the time of the photograph, it was one of two (GATX 26879 being the other) listed together. Compare this car to AUPX 102 on page 24.

(Paul C. Winters)

GATX 29277
AAR203W
▲ Note the Anheuser Busch logo on the dome in this July 1968 illustration. It was assembled by General American in March 1958. A January 1968 equipment register places the car within the GATX group 29271-29280. *(Emery Gulash, Morning Sun Books collection)*

GATX 38142
DOT103AW

▼ We're looking at a car found in Tullahoma, Tennessee on March 21, 1975. It was built at General American in January 1961. *(Lee Yoder)*

GATX 54897
DOT103ALW

◄ This tank car, with its aluminum tank, was photographed at Elizabethport, New Jersey in April 1972. It was manufactured in January 1952. Rubber lined it was listed as part of the 20-car set GATX 54880-54899.

(William Rosenberg)

GATX 59589
ICC103AW

▼ Constructed by General American in December 1948, it sits in June 1965. By April 1949 it was tabulated as part of the 92-car series 59508-59599 used in acid service. It remains to be seen whether all cars were similar.

(Bernie Wooller)

GATX 60919
ICC103AW
▲ Another car under lease to Olin Mathieson, but constructed by General American in March 1950. This one was put on film in December 1965 and was described as one of 102 in the group 60843-60944 of acid service cars in October 1950. A close look at the bolsters will show poling pockets set within them. *(Bernie Wooller)*

GATX 64658
ICC103BW
▲ Yet another acid service car. It was most likely photographed at St. Louis in September 1967. The car was constructed by General American in February 1954 and by January 1955 was listed as within the 64619-64699 number set. *(Paul C. Winters)*

GATX 64693
ICC103BW
◄ A strange thing here. While in much nicer paint than the car above, and from the same series, note that this car has different dome and tank saddle designs. Yet its manufacture date is exactly the same as the car above. Obviously at least two sets of cars were involved with this group. The photo was taken at Pensacola, Florida on May 11, 1986.

(Emery Gulash, Morning Sun Books collection)

GATX 69517
ICC111A100W5

◀ This car does not have an aluminum tank since its tank is made of steel: it has simply been painted so. It was assembled by General American in November 1950 and was rubber lined. As built it belonged to the 69500-69599 number series. But based on available evidence it is thought that the car was one of three (69515-69517) converted from the original AAR205A300W specification at the time of this April 1962 photo taken in Columbus, Ohio. *(Paul C. Winters)*

GATX 69597
ICC109A300W

▼ General American constructed this car in January 1951 and was part of the series described above. It was spotted at Lafeyette, Indiana in April 1960 hauling ammoniated nitrate solution which was utilized by the car's lessee to make specialized industrial coatings, "Spensol" being one tradename. Also like the car above, it is likely converted from its original AAR205A300W specification. *(Lee Yoder)*

GATX 71168
ICC103CW

▲ Photographed in October 1962 at the General American plant at Sharon, Pennsylvania, it is believed to have been constructed that February. As such, it was part of the GATX series 71143-71182 of cars equipped for acid service. This is another car with a steel tank painted silver.

(Calvin T. Banse)

GATX 73152
ICC103W
▶ Although this car looks quite similar to the Anheuser Busch car shown on page 29, take note of the subtle lettering differences. Its photo was taken in August 1967 and the car was constructed in April 1958. By July of that year it was placed within the 73100-73199 number set. *(William Rosenberg)*

GATX 74789
▼ A car put on film in July 1958 and under lease by a large manufacturer of industrial chemicals. Apparently constructed in 1952, by January 1953 it was placed within the 74700-74799 set of 100 cars.
(Emery Gulash, Morning Sun Books collection)

GATX 75450
ICC103W
▼ A car assembled by General American in September 1957 and photographed in February 1964. *(Paul C. Winters)*

GATX 80327
DOT103W

▲ A picture taken in August 1971. About the only thing known is that at the time of its photo it was fitted somewhere within the 80140-80336 group. The lessee of this car needs no introduction, of course!
(Paul C. Winters)

GATX 81980
ICC103W

▼ Photographed in May 1976 the car was at Amana, Iowa and leased by a corn products manufacturer. Its manufacture date is unknown, but in January 1968 it was one of 31 in the 81957-81987 group. The car is equipped with heater coils. *(Gib Allbach)*

GATX 83213
ICC105A300W

▶ Photographed at Sharon, Pennsylvania in October 1962, the car appears to have markings indicating that it had been fabricated in January of that year. Between then and early 1964 it was shown as one of seven individual cars placed at random throughout the 83200-83399 number series. The company leasing this car was a wholesale propane company.
(Calvin T. Banse, Morning Sun Books collection)

GATX 87325
ICC111A100W
▼ Built January 1960 it sits at Columbus, Ohio in April 1962. Equipped with rather unusual band rings for the hand rail supports, it was one of 26 in the GATX 87324-87349 number set. *(Paul C. Winters)*

GATX 95163
DOT111A100W2
◀ At the time of its photograph (taken in December 1975) it was one of 82 cars in acid service scattered within the larger series 93968-96050. It is impossible to say if all cars were similar to this one. *(William Rosenberg)*

HCPX 1758
ICC103W
▼ As can be seen, these reporting marks were assigned to the Hooker Chemical Corporation. Photographed in February 1981 it was part of the 1758-1761 set of 8,000 gallon tank cars. The car was built in October 1966 by American Car and Foundry's Milton plant.
(Rail Data Service, Jim Kinkaid collection)

NATX 19223
▶ Found in August 1967, this is an 8,055 gallon capacity car constructed about July 1961. It was part of the NATX group of cars numbered 19200-19265. Rohm and Haas was a large company (begun in 1907 and operated by both American and German owners) that made various speciality materials and chemicals. *(William Rosenberg)*

NATX 20131
DOT111A60ALW
▼ A car constructed by AMF Beaird in August 1966. It is an interesting car with its dual anchor rods at each corner and four reinforcement tank rings! At Birmingham, Alabama in May 1971, the car was part of the 20129-20139 set. *(Jim Thorington)*

NATX 21593
ICC111A100W1
▲ Another car under lease to Rohm and Haas, but it is much larger, being of 20,715 gallon capacity. The photo was taken in Birmingham, Alabama in July 1964. The tank for this car was built by AMF in May 1964 and the car was assembled by North American's Texarkana, Texas plant that same month. It was one of 101 in the North American series 21531-21631. *(Jim Thorington)*

NATX 21715
DOT111A100W1

▲ This shot was taken at Altoona, Pennsylvania in June 1990. While plainly painted, it is interesting nonetheless due to its walkway and dual tie downs at each end. The car was assembled in December 1964 as part of the NATX series 21632-21720. *(Jim Kinkaid)*

NATX 25104
ICC111A100W5

▶ Another rather unusual car found in February 1966. It was one of 20 assembled in December 1965 by North American's Texarkana shops and was part of the 25103-25122 number group. ERCO Chemicals, in business since 1897, was the largest producer of sodium chlorate (a herbicide) in North America. *(Bernie Wooller)*

PLCX 180007
DOT111A60ALW2

▼ A builder's photo of an aluminum tanked car built for hydrogen peroxide service. It was constructed in December 1987 at Trinity Industries' Tulsa, Oklahoma plant and sold to Pullman-Standard's leasing division, who in turn leased it to the FMC Corporation. An 8,218 gallon car, its series was 180000-180009.

(Pullman-Standard photo, Jim Kinkaid collection)

PQX 850
ICC103
▼ PQX was the reporting mark for the Philadelphia Quartz Company, founded in 1831. The car shown here was built by ACF as part of the PQX series 849-858. Fitted with steam coils it was photographed in June 1978 in Birmingham, Alabama. This set was assembled under lot 1954 in late 1939 or early 1940 at the Milton plant and were 8,000 gallon cars. The product carried, soluble silicates, is utilized in many manufacturing processes ranging from paper to tires. *(Jim Thorington)*

SCHX 28101
DOT103W
◄ The Stauffer Chemical Company underwent a general reorganization in early 1956 resulting in the cars operated by the various branches being consolidated into two reporting marks, of which this was one. Here's a car built in October 1954 and photographed on August 26, 1973. It was part of the 28101-28111 group. The source for this set remains unknown.
(William Rosenberg)

SCHX 41105
ICC103W
▼ Another Stauffer Chemical car assembled in October 1957. It was photographed in September 1970. The car belongs to the SCHX series 41101-41105, which were 10,000 gallon cars.
(Rail Data Service, Jim Kinkaid collection)

SCJX 77516
DOT111A100W1
▲ This reporting mark was originally used by the Spencer Chemical Company who was originally a maker of military explosives. In 1963 it was bought by Gulf Oil and in early 1965 the reporting marks were transferred to the Warren Petroleum Company, a subsidiary of Gulf Oil. The car was constructed by General American in February 1953. However, the series that this car belongs to (SCJX 77500-77549) did not appear until July 1965. These cars were never marked SCJX and had been acquired from another source. Our example was found in July 1973. *(Paul C. Winters)*

SHPX 6936 & 7793
ICC103AW & ICC103W
◀▼ Two cars leased by Penn Salt Chemicals. Penn Salt, originally named the Pennsylvania Salt Manufacturing Company, has existed since at least 1850. In 1969, due to a merger, it was renamed the Pennwalt Company. At middle is a car photographed at Columbus, Ohio in November 1962. Its markings indicate a possible June 1954 manufacture date, which matches its first listing in the October 1954 equipment register within the group 6926-6940. Since it carries hydrofluoric acid, it was probably assigned to the Calvert City, Kentucky plant. Below this car is another one photographed in May 1968. Rubber lined, it was built in July 1958 at ACF's Milton, Pennsylvania plant. The series that the car belongs to would appear to be SHPX 7791-7794. The product that this car is carrying is used primarily in water treatment. *(both, Paul C. Winters)*

SHPX 12327
ICC111A100W1

▼ This photo was taken in December 1963. The car was assembled in February of that year at Milton and has steam coils and was part of the 12300-12339 number group. Borden Chemical, producer of the familiar glues and resins, was the result of a 1958 renaming of the chemical side of the business. *(Paul C. Winters)*

SHPX 12493
ICC111A100W1

▲ Fabricated in February 1962 this car was sitting in Columbus, Ohio in July 1962 when put on film. It was constructed as part of the set SHPX 12488-12499. *(Paul C. Winters)*

SHPX 19703
ICC103A-ALW

▲ Here is a car built by ACF in September 1955 with an aluminum tank. It was part of a five car set SHPX 19700-19704. When it was photographed at Council Bluffs, Iowa on March 5, 1972 it was carrying hydrogen peroxide. *(Lou Schmitz)*

SHPX 22813
ICC103W
▲ A car assembled by American Car and Foundry in April 1952 as one of five in the series 22813-22817. Built under lot 02-3648A, it was an 8,088 gallon coil equipped car.

(Emery Gulash, Morning Sun Books collection)

SHPX 24480
ICC103W
▼ Part of a 25-car order from ACF, it was assembled under lot 72-5715 in June 1961. Here the 8,106 gallon car, equipped with coils, was at Columbus, Ohio in April 1962. Union Starch and Refining was an established corn processing company and they leased all of the cars in this order, which were numbered 24458-24482. *(Paul C. Winters)*

TELX 17076
DOT111S60ALW2
▼ ELF Atochem (a company formed in 1984 to make industrial and speciality chemicals) operated this car, found at its Memphis, Tennessee home point in December 2000. It was built in January 1985 and from the 14-car series 17060-17078. This series first appeared in April 1999 and the cars were acquired used from the DuPont company (see page 26 for one in DuPont livery) since ELF Atochem acquired the Memphis hydrogen peroxide plant from them in 1998.

(Al Lanier, Jim Kinkaid collection)

TVAX 404
DOT103CW
▲ The Tennessee Valley Authority, who was assigned these reporting marks, has had cars listed since January 1963. Yet this car, which appears to have been constructed in February 1942, only first showed up in the register in October 1979 as one of three in the 403-405 series. The 7,500 gallon car was used in liquid fertilizer service and was in Birmingham, Alabama when photographed in March 1980. *(Jim Thorington)*

UCLX 801
ICC103BW
▼ Retired and awaiting scrapping by its owner, Vulcan Chemicals, it sits near the Vulcan Plant outside of Wichita, Kansas on March 30, 1991. The reporting mark originally stood for Union Chemical Car Line. The rubber lined car was fabricated in October 1955 for muriatic acid use and was part of the UCLX 801-809 set. *(Jim Kinkaid)*

UTLX 13220
DOT111A100W2
▲ Sitting at Birmingham, Alabama it was caught by the camera on July 19, 1981. Obviously the tank bands need some tightening! Lessee Grace Davison (part of W.R. Grace) was a speciality chemical manufacturer and in this instance was hauling a substance called "rare earth chloride," used primarily in the steel industry. The series that the 10,840 gallon car belonged to is thought to have been UTLX 13216-13220.

(Al Lanier, Jim Kinkaid collection)

UTLX 14378
ICC103BW

▶ As can be seen the car was constructed in September 1955 by Union Tank Car. It was photographed in February 1968 and was rubber lined. Documents suggest that it was part of the UTLX series 14378-14398, which were 8,187 gallon cars. *(Bernie Wooller)*

UTLX 47279
ICC103W

◀ Although in plain black, observe how the car was repainted. The shop simply painted around the smaller lettering on the right hand side of the tank and didn't even bother to finish the car! This 8,100 gallon car was assembled in March 1958 and was under lease to the American Oil Company when photographed in November 1972. It appears to have heater coils and in 1958 was assigned to the 47088-47287 group.
(William Rosenberg)

UTLX 59055
AAR203W

▼ At Birmingham, Alabama the 8,050 gallon car was photographed in November 1975. It displays a June 1957 build date and is likely from the UTLX series 59050-59072 group. This would seem to be another example of a car that needs its tank bands tightened up some.
(Jim Thorington)

UTLX 86058
ICC103ALW
▲ Although similar to UTLX 86026 on the previous page, it displays an April 1964 build date. Carrying the commodity toxaphene, an insecticide, the car was photographed in November 1964.
(Jim Thorington)

UTLX 86879
▶ At the time of its photo, taken in April 1969, the car was placed with others in the series 86850-86879.
(William Rosenberg)

WRNX 20006
DOT111A100W1
▼ Our last non-insulated car is this photo taken in November 1969 of a car constructed by General American in April 1962. But it wasn't until January 1963 that Warren Petroleum began listing the car. It was a part of the 20000-20008 number group. This was a 20,427 gallon capacity car fitted with heater coils.
(Paul C. Winters)

Insulated Single Compartment Cars

AAMX 5157
None stenciled

▶ Our first insulated car happens to be from south of the American border. The reporting marks were first used by the by the Mexican company Arrendadora de Carros de Ferrocarril del Atlantico, S.A. in October 1974. The car, leased by Liquid Carbonic de Mexico, S.A. and photographed at Los Mochis on March 14, 1975, made its first register appearance in January 1975 as a single car entry. It carries liquefied carbon dioxide. *(Matt Herson)*

ACDX 49550
AAR201A80W

▼ Assembled in May 1951 by General American Transportation Company and of 8,085 gallon capacity, the car was caught on film in June 1974. It was one of 110 cars constructed as SPX 9500-9609 for the Solvay Process Division of Allied Chemical. By January 1953 the entire set had been transferred to the Nitrogen Division and given NDX reporting marks, with no number change. Along the way, an extra seven cars were added to the total. In April 1963, when all of Allied Chemical's fleet was combined, the entire series was renumbered as ACDX 49500-49616. *(Bernie Wooller)*

ACDX 83000
DOT103W

◀ Fabricated by American Car and Foundry, this example was built in May 1951 under lot 3498. As such, it was one of three cars supplied as BMX (reporting marks for the Barrett Division of Allied Chemicals, who made products from coal tar for such uses as dyestuffs and drugs) 3000-3002. In late 1958 the cars were transferred to PLCX reporting marks, keeping their numbers. As the various Allied Chemical cars were folded into one register listing (now called First Union Properties) in April 1963, the 3 cars were added with 11 others and renumbered as ACDX 83000-83013. It was put on film in July 1972.

(William Rosenberg)

ACDX 89145
▶ With origins from BMX 9100-9199, it was constructed by General American. They were then lumped together with BMX 9009-9099, constructed by American Car and Foundry. The group was interesting in that the series was mixed between two Allied Chemical divisions: some kept BMX reporting marks while others got PLCX markings. Milwaukee, Wisconsin was the place of this photograph.
(John Luckfield, Jim Kinkaid collection)

ACDX 812539
DOT103W
▼ Photographed in March 1971 it was assembled by American Car and Foundry in December 1952. The car was one of 100 in the BMX group 12500-12599 and fabricated under lot 02-3920. These 12,500 gallon cars were originally supplied for asphalt service. *(William Rosenberg)*

ACFX 3745
ICC105A500W
▶ On March 22, 1984 this car was photographed at Calvert City, Kentucky where lessee Pennwalt (formerly Penn Salt, see page 39) made hydrofluoric acid. It had originally been built in March 1952 as part of the SHPX 3737-3748 group of 10,500 gallon chlorine cars. They were assembled by ACF under lot 02-3636E.
(Al Lanier, Jim Kinkaid collection)

ACFX 3771
ICC105A400W

▲ Caught on film in June 1969 the car was built as part of the SHPX group 3750-3799 in late 1948. American Car and Foundry assembled these cars under lot 3289. The company stenciled on the car was a result of an August 1954 merger between Olin Industries and Mathieson Chemical. This new company manufactured such things as chlorine, caustic soda and polyurethane foams. *(Paul C. Winters)*

ACFX 4948
DOT105A300W

▼ With rather unusual markings it sits at Memphis, Tennessee on August 13, 1981 while under lease to the DuPont Company. Carrying hydrocyanic acid (note the poison gas placard) the placard board gives instructions as to who to call in emergencies. The car was probably part of the ACFX 4943-4952 group, built in August 1957 originally for chlorine service. *(Al Lanier, Jim Kinkaid collection)*

ACFX 6275
ICC103W

◀ With a 6,123 gallon capacity and photographed in February 1973 it would appear to be one of 4 originally marked as SHPX 6272-6275. The cars were assembled in late 1961. Besides starches National Starch and Chemical, in business since 1959, also supplied adhesives.

(William Rosenberg)

ACFX 15166
ICC103W
▼ Photographed in May 1969 it had been assembled by American Car and Foundry in November 1946 under lot 2965. The car belonged to the original SHPX set 15140-15174 which were used for caustic soda service. Wyandotte Chemicals (later BASF) was a manufacturer of soaps and cleaners. *(Jim Thorington)*

ACFX 19121
◀ Photographed in March 1971 at Trenton, New Jersey. Manufactured for chlorine service the car is believed to be part of the 143-car series 19000-19142, assembled in mid-1965 and early 1966. The lessee of this car (GAF Corporation, renamed from General Aniline and Film in April 1968) was a supplier of dyestuffs and chemicals.
(William Rosenberg)

ACFX 85022
ICC105A300W
▼ While a rather plain car, as the photo taken in April 1969 shows, this large car (in refrigerant gas service) indeed has a full length center sill. It was manufactured by American Car and Foundry in May 1964 and the original series is believed to have been SHPX 85000-85039.
(Paul C. Winters)

ACFX 90240
ICC111A100W1
▲ Photographed on May 6, 1972 its markings say that it was manufactured by American Car and Foundry in June 1943. Because the 10 car series that this one belonged to (ACFX 90236-90245) was first listed in October 1964, it's possible that these cars were renumbered. The car in this illustration carried 10,999 gallons of paint! *(William Rosenberg)*

ADMX 43028
DOT105A500W
▶ Archer-Daniels-Midland operated 25 of these cars numbered 43026-43050. One such example sits in Kansas City on April 3, 1987, having been constructed in March 1986. The set was assembled by Trinity Industries at Tulsa, Oklahoma, and construction carried over to April.

(Jim Kinkaid collection)

AESX 4936
◀ North American Car assembled this car at its Texarkana, Texas plant in February 1964. The 8,419 gallon car was one of 126 in the AESX group 4900-5025, first listed in the January 1964 issue of the equipment register. This one was photographed in August 1967 in corn syrup service. *(William Rosenberg)*

ARIX 1123
DOT015A500W
▶ Photographed in July 1972 the 18,248 gallon capacity car was rostered by the Airco Industrial Gases Division of the Air Reduction Company. The car was first listed in the equipment register as one of 20 in the 1114-1133 group in October 1967.
(William Rosenberg)

ASRX 1653
DOT111A100W1
▼ The American Sugar Refining Company (later the American Sugar Company) operated this car, which was one of five in the 1651-1655 set. Although the car is marked as having been constructed by ACF in February 1953, it's thought that it was actually assembled in February 1963 (perhaps by General American) since the first equipment register listing was in January 1963. At Baltimore, Maryland in August 1985. *(Jim Rogers)*

CCBX 2398
ICC111A60W1
▲ Assembled by ACF in December 1944 it first appeared in the July 1965 equipment register as one of four marked CCBX 2396-2399. The 11,000 gallon car had heater coils. Because this specification did not exist in 1944 it is suspected that owner Union Carbide rebuilt them from older cars. *(Paul C. Winters)*

CCLX 294
DOT203W

▼ The Crystal Car Line, who operated cars under this reporting mark for the Corn Products Company, (later CPC International) had a number of various paint schemes over the years. Corn Products transported a number of products derived from corn (naturally) used primarily in the food industry. The string illustrated here was photographed in May 1971. Car 294 would appear to be one of 135 introduced in January 1963 and numbered as 201-335. *(William Rosenberg)*

CCLX 443
DOT203W

▲ Assembled by General American Transportation in October 1958, it was one of 100 in the series 400-499. These 8,000 gallon cars first appeared with the October 1958 equipment register. *(Paul C. Winters)*

CCLX 809
DOT203W

▼ Photographed in October 1966 it belongs to the CCLX series 800-839. That group was first listed in the October 1957 equipment register. *(William Rosenberg)*

CCLX 991
▲ DOT203W
At Trenton, New Jersey in May 1972. Assembled by ACF in mid-1958 it was first mentioned in the equipment register in October of that year. The group that it was part of was CCLX 975-999, all 8,000 gallon capacity cars. *(William Rosenberg)*

CCLX 1092 & 1110
DOT103W
▲▼ A couple of cars from the 100-car set 1021-1120 built in early 1964. Car 1092 was photographed in October 1974 and was assembled by General American Transportation. Meanwhile, yet another paint scheme is shown as tank car 1110 travels through Kansas City, Missouri on March 25, 1992. *(middle, Paul C. Winters; bottom, Jim Kinkaid)*

CCLX 3057
DOT103W
▲ Our last Crystal Car Line tank car sits in Birmingham, Alabama in March 1971. It was assembled by General American in April 1964. The car was part of a 55-car set numbered CCLX 3042-3096. *(Jim Thorington)*

CELX 1050
◄ Celtran (the transportation subsidiary of the Celanese Corporation) owns this reporting mark. This example moves through Tullahoma, Tennessee in May 2005. The car was part of the 1000-1055 set which had been assembled by Richmond Tank Car of Houston, Texas in October 1964 through January 1965. It carries acetic acid, used to make vinyl acetate (an intermediate chemical used in a wide array of manufacturing processes). *(Lee Yoder)*

CELX 1123
DOT111A100ALW1
▲ Another car constructed by Richmond Tank Car. Found at Altoona, Pennsylvania in June 1990 it is from the 1100-1139 group, which carried both acetic acid and ethyl acetate (used as a paint solvent). Close inspection will show that the left hand truck has two sizes of wheels installed! These cars were assembled from December 1966 through February 1967. *(Jim Kinkaid)*

CNCX 20017
ICC103ALW

◀ We are in Birmingham, Alabama in April 1970. American Car and Foundry assembled the car in January 1964 as part of the 20-car set CNCX 20001-20020. Columbia Nitrogen, founded in 1962, was involved with nitrogen-based fertilizers. *(Jim Thorington)*

DAX 740
DOT105A500W

▼ A 10,818 gallon car sits at Milwaukee, Wisconsin in April 1975. The car was part of the DAX series 731-745, a 15-car set that first appeared in early 1957. Diamond Shamrock was the product of a 1967 merger between Diamond Alkali (and from whence the reporting mark originally derived) and Shamrock Oil and Gas.

(John Luckfield, Jim Kinkaid collection)

DAX 1292
DOT103W

▶ A car put on film in July 1974 that had been fabricated in January 1956 by General American Transportation. It was part of the 100-car group numbered 1200-1299. The car illustrated here was employed in caustic soda service but Diamond Shamrock was also involved with the oil and petrochemical industries. *(Bernie Wooller)*

DOWX 3604
DOT111A60W1

▼ DOWX, naturally enough, is assigned to the Dow Company as can be seen here. Our example was at Newark, New Jersey in May 1969 and was supplied by ACF's Milton shops in December 1966. It was one of 21 cars assigned to the DOWX 3600-3620 series. Fitted with heater coils they were leased via the "Commonwealth Plan," a financial instrument of the Bankers Leasing Company (which also furnished cars to the MKT and SP railroads along with privately marked cars).

(William Rosenberg)

DUPX 6088
DOT105A300W

▲ The DuPont Company had a large and varying fleet of tank cars, many quite colorful and unusual. One example sits in Wellington, Kansas in December 1989. Motor fuel anti-knock compound is fairly dense, accounting for the smaller tank size. It is likely part of the series 6060-6109 assembled in mid-1956. *(Jim Kinkaid)*

DUPX 6553
DOT103W

▼ One passes through Altoona, Pennsylvania in June 1990. Here is a car assembled by General American Transportation in late 1957 and presumably part of the 6519-6579 set. Of interest here is the wrinkled and patched outer jacket on the right end of the car. *(Jim Kinkaid)*

DUPX 8064
DOT105A300W

▼ A car in Memphis, Tennessee on October 3, 1983. Assembled in July 1956, field notes conflict here: it is thought that the car was part of the DUPX group 8051-8070, though the builder remains in question. *(Al Lanier, Jim Kinkaid collection)*

DUPX 8361
ICC105A500W

▼ Fabricated by American Car and Foundry in May 1957, this DuPont car was photographed by the United States Railway Equipment Company at Blue Island, Illinois in May 1963. It was constructed under lot 02-4960 as part of the DUPX series 8361-8370, all for Freon© service.

(USRE photo, Jim Kinkaid collection)

DUPX 8361
DOT120A500W

▶ Twenty six years later the same car sits at Centralia, Illinois in September 1989. Besides the much plainer paint scheme, note that the car has been converted to another specification. One nice thing about plain white cars: it is easy enough to see all the construction details!
(Jim Kinkaid)

DUPX 8419 & 8424
DOT105A300W

◄▼ Two hydrocyanic acid cars. At top is one found in February 1967. Below that is another found in February 1977. Both cars were assembled in April 1958. Although the builder is unknown the series was DUPX 8401-8432.

(top, Jim Thorington; middle, Paul C. Winters)

DUPX 8505
ICC105A300W

▲ In March 1962 this natty car was photographed in Columbus, Ohio in the Pennsylvania yards at St. Clair Avenue. It was in isocyanate acid service and was constructed in February 1961 by American Car and Foundry. The welded structure to the left of the manway platform is presumably some kind of guard for the relief valve. The car was part of the DUPX 8501-8505 group. *(Paul C. Winters)*

DUPX 8829
ICC105A300W

▲ In May 1962, also at Columbus, Ohio but at the Pennsylvania's Grandview Yard. Assembled in May 1957 by American Car and Foundry for metallic sodium service it was part of the DUPX 8824-8843 group, constructed under lot 02-4955. At the time, metallic sodium's major use was in the manufacture of tetra-ethyl lead, a motor fuel antiknock compund.

(Paul C. Winters)

DUPX 9926
DOT105A300W

◄ A car placed on film in August 1972. It is thought that the 9,000 gallon car was one of 83 in the DuPont series 9901-9983.

(William Rosenberg)

DUPX 20016
DOT105A300W

▼ Built by General American Transportation in October 1960 it is in anhydrous hydrofluoric acid service. This photo was taken in June 1962 and the car was one of 18 in the DUPX group 20001-20018.

(Emery Gulash, Morning Sun Books collection)

EBAX 119
AAR203W
▲ These reporting marks were assigned to the Ethyl Corporation, a maker of automotive fuel anti-knock additives. EBAX 107-121 (odd numbers only) first appeared in the July 1970 equipment register. As this particular car was manufactured in April 1953 by General American Transportation the cars were probably acquired used. It was photographed at Gibsland, Alabama on January 17, 1982. *(Al Lanier, Jim Kinkaid collection)*

EBAX 408
DOT105A300W
▲ Here is an 11,000 gallon tank car from the series 406-409. It was built in March 1944 and photographed in June 1968. *(Paul C. Winters)*

EBAX 1006
DOT105A300W
▲ Richmond Tank Car assembled the car shown here in July 1968. As such, it was part of the 10-car group numbered 1000-1009. Ethyl 1006 was spotted in the Milwaukee, Wisconsin area in August 1976. *(Don Reck, Gib Allbach collection)*

EBAX 6020
ICC105A300W
▲ This picture was taken in May 1968. The car first appeared in early 1949 as part of the 6012-6023 series and were of 6,000 gallon nominal capacity. General American Transportation Company assembled these cars. *(Paul C. Winters)*

EBAX 6258
ICC105A300W
▶ American Car and Foundry fabricated the car illustrated here in April 1956. It was part of the 6,000 gallon capacity group numbered 6235-6258, assembled under builder's lot 02-4679. Ethyl 6258 was photographed in May 1969. *(William Rosenberg)*

EBAX 6263
DOT105A300W
▼ A photo taken on May 13, 1972 at Flat Rock, Michigan of a car fabricated by General American Transportation in April 1956. It is part of the EBAX group 6259-6282. Note that this set and the ACF-supplied series just above were both combined into one series in the equipment register.

(Dave McKay, Morning Sun Books collection)

GATX 8186
▲ At Birmingham, Alabama in March 1969. Possibly assembled by General American in late 1966 or early 1967, it was one of 20 first listed in the July 1967 register as GATX 8175-8194. The lessee of this car was involved with nonmetallic minerals such as kaolin and fullers earth.
(Jim Thorington)

GATX 11528
DOT105A300W
▶ Assembled by General American in March 1970, it was caught on film in April 1971. Lead lined it was one of three in the GATX group 11527-11529. Great Lakes Chemical sold the bromine that this car was carrying for use in the fire suppression industry and the material was also utilized for water purification.
(William Rosenberg)

GATX 27909
DOT105A500W
◀ Built by General American in March 1970 for carbon dioxide service it was photographed one year later, in March 1972. A 19,795 gallon capacity car it belonged to the 20-car set 27900-27919.
(William Rosenberg)

61

GATX 27919
DOT105A500W

▶ Another carbon dioxide car from GATX 27900-27919. Note that is has a different Liquid Carbonic (who shipped carloads of this material for industrial gas use) paint scheme than GATX 27909 on the previous page. At Sterling, Ohio on June 6, 1976 it would appear to have just been released from the Sharon paint shop.

(Dave McKay, Morning Sun Books collection)

GATX 32193
ICC105A500W

▼ Assembled in August 1948, it was photographed in April 1960 while in chlorine service. By April 1949 the equipment register placed this car within the 32100-32399 series, though only 203 cars were accounted for. *(Emery Gulash, Morning Sun Books collection)*

GATX 32248
◀ It was photographed in January 1955 and would appear to have been one of 84 in the GATX group 32216-32299, which were first introduced in the register circa 1947. The September 1948 date on the car's side may be its construction date. The car was carrying ethylene oxide, which Jefferson Chemical manufactured for use in making antifreeze. It's also used to sterilize medical products.

(Emery Gulash, Morning Sun Books collection)

GATX 32835
ICC103W

◀ A car photographed in May 1968, probably in Birmingham, Alabama. As can be seen, it was supplied by General American in April 1941. By October 1950 it was shown as part of the 32830-32838 set. Caustic soda, which this car is employed in carrying, is used in numerous applications, including paper, pulp and aluminum manufacture, along with food preparation use. *(Jim Thorington)*

GATX 33656
AAR203W

▲ In a photo taken in May 1967 we see a car constructed by General American in July 1948. The fittings at each end of the tank appear to be cleanout ports. There were 52 of these cars in the GATX group 33648-33699 by April 1949.
(Emery Gulash, Morning Sun Books collection)

GATX 33778
ICC105A500W

▼ Vulcan Chemicals leased this car for liquid sulfur dioxide service. This stuff was utilized to make such diverse things as sulfuric acid (for automobile batteries for example) and wine! It was at Tullahoma, Tennessee when photographed in June 1978. General American constructed this car in January 1964 and was apparently part of the 33760-33790 set of 31 cars. *(Lee Yoder)*

GATX 33988
DOT105A500W

◀ In chlorine service we see the car as it was painted in March 1971. It was fabricated by General American in June 1966 as part of the number series 33965-33999.
(William Rosenberg)

GATX 34445

▼ A rather unusual car, who's photo was taken in March 1975. It was one of six acid service cars in the set 34443-34448. The first of those cars appeared in January 1958 and the full six were listed by April 1959. The reason for the full length walkway is quite unknown.
(Paul C. Winters)

GATX 36268
ICC111A60W1

▲ General American built this car in January 1947. It was at Birmingham, Alabama in September 1964 when it was caught on film. It may have been originally constructed as a pressure car and converted in 1963 to general service. *(Jim Thorington)*

GATX 36372
ICC111A100W1
▼ Another shot taken at Birmingham, Alabama, but in June 1964. An 8,245 gallon car constructed in August 1948, it may have been converted at the Argentine, Kansas shops in October 1959. By January 1960 it was one of six listed with various numbers. The lessee was a company based out of New Orleans, Louisiana and was named for founder Leon Godchaux. *(Jim Thorington)*

GATX 38921
ICC103W
▲ We're looking at a car in phosphorus service, a product that is useful in metal and water treatment and in pesticides. It is carrying markings indicating that it was fabricated in March 1960 and was photographed in April 1969. General American placed this car somewhere within the 38850-38999 number group. Again, note the cleanout access ports on each end. *(Paul C. Winters)*

GATX 61175
ICC103W
◀ In January 1955 the photographer caught this car on film. It was built by General American in September 1953 and was likely part of the three car set 61173-61175. The Solvay Process marked on the car refers to a method of making soda ash, which has numerous manufacturing uses such as glass, soap and paper manufacturing.

(Emery Gulash, Morning Sun Books collection)

GATX 62511
ICC103W
▶ This car is impressive with its aluminum-painted tank jacket. It was found at General American's Sharon, Pennsylvania plant in October 1962. Built in June 1948 it may have been part of the set 62400-62567.
(Calvin T. Banse, Morning Sun Books collection)

GATX 63210
ICC103W
▼ Assembled by General American in September 1961 the car was part of the GATX group 63100-63299. It was caught on film in February 1972.
(Dave Nelson, David Hickcox collection)

GATX 63718
ICC105A300W
◀ A car found scooting along in November 1967. It is hard to tell, but it appears that it was assembled by General American in March 1949. The company placed this car somewhere within the 63600-63899 series.
(William Rosenberg)

GATX 63837
ICC105A300W
▲ Found in August 1978 at Ft. Meade Junction, Missouri hauling motor fuel anti-knock compound, it was assembled in April 1953. Cork insulated, the car carried slightly more than 6,000 gallons of cargo. Note that even though of different design than 63718 on the previous page, it was part of the same overall number group suggesting multiple orders were involved in the series. *(Jim Rogers)*

GATX 63856
ICC105A300W
▲ Carrying liquefied petroleum gas when photographed in April 1963 it was one of 126 cars in the number group 63854-63979. There are no visible dates on this car to indicate when it was built but these cars are likely to have been renumbered from the original Warren series WRNX 2000-2095 and 2245-2275 circa 1962. *(Paul C. Winters)*

GATX 64548
ICC105A300W
◄ An example at Birmingham, Alabama in March 1969. It is employed in anhydrous ammonia service and was constructed by General American in September 1951. The register placed this car within the GATX set 64546-64553. Perhaps the largest use of anhydrous ammonia is for agricultural use. *(Jim Thorington)*

GATX 66980
ICC105A300W
▼ At Missouri Pacific's Wichita, Kansas yard in November 1989. The six cars in this series (GATX 66980-66985) were assembled by General American in May and June 1969. Chemetron (an acronym for "Chemical-Metals-Electronics") and Cardox, who merged with Chemetron in 1958, ship this car's commodity as part of their industrial fire control agent business. *(Jim Kinkaid)*

GATX 67626
ICC105A400W
▲ Built by General American in November 1949, here is a car photographed in May 1968. Union Carbide's trademarked "Ucon" fluorocarbons are used primarily as aerosol propellants.
(Paul C. Winters)

GATX 70840
DOT111A60ALW1
▲ A car found in Dayton, Ohio in September 1989. The lessee of this car was a speciality chemical manufacturer based in Cincinnati, Ohio where they had been in business since 1840. *(Jim Kinkaid)*

GATX 73339 & 73346
ICC103W
▶▼ Two cars carrying caustic soda. The top photo was taken in December 1968. The middle illustration was taken on February 25, 1962 at the Pennsylvania Railroad's St. Clair yard hump in Columbus, Ohio. It is hard to state with certainty when the cars were constructed, but the register showed them to be part of GATX 73303-73348 as of January 1961.
(both, Paul C. Winters)

GATX 78848
DOT111A60W1
▲ Built in May 1952 this car sits for its portrait in April 1971. Of interest is that weld line on the jacket head: note how it slopes down towards the far side of the car! The company J.M. Huber (founded in 1890 by Joseph M. Huber) is a speciality additive producer that makes products used in everything from paper and tire manufacturing to toothpaste. *(William Rosenberg)*

GATX 79415
DOT105A300W

◀ The photo was taken in Memphis, Tennessee in May 1985. At that point in time, it was one of 22 variously numbered within the 79309-79484 number set. The commodity carried in this car is basically hydrogen cyanide in water (also known as Prussic acid). This is utilized as an intermediate chemical to make other chemicals used primarily in the mining industry.
(Al Lanier, Jim Kinkaid collection)

GATX 79469
▼ A car put on film in May 1971. *(William Rosenberg)*

GATX 80093
DOT111A60W1
▼ Found in June 1973, at that point in time it was simply listed as one of 25 cars within the 80000-80097 number set. It had been constructed by ACF in August 1952. Quaker Oat's Chemical Division (begun in 1921) produced furfural in Memphis, Tennessee. Furfural was mostly used in lubrication oil refining. *(David Nelson, David Hickcox collection)*

GATX 80489
ICC105A100W
▲ The history on this car, photographed in October 1974, is a bit murky. It is thought to be part of the GATX group 80471-80500, constructed about March 1960 as 11,080 gallon capacity cars. It was being leased by a refinery based in Channelview, Texas. *(Paul C. Winters)*

GATX 81188
ICC103W
▲ Here's a photo taken in August 1964. The car was built that month by General American and was part of a string of similar cars underway, probably en route towards their lessee for their first load of rosin (a product derived from kraft paper pulping and used in finished paper manufacturing). For this service it carried heater pipes. The 9,970 gallon car was part of the 81178-81199 number series.

(Paul C. Winters)

GATX 83507
DOT105A500W
◄ Philadelphia, Pennsylvania was the location of this photo, which was taken on July 9, 1980. It is believed that the 17,265 gallon car was part of the 8-car set numbered as GATX 83506-83513, assembled in August and September 1962.

(Al Lanier, Jim Kinkaid collection)

GATX 84597

▶ A car that was at Birmingham, Alabama in May 1960. As it is being leased by the Thiele Kaolin Company (of Sandersville, Georgia) it is likely carrying kaolin, a mineral being transported as a slurry generally used for finished paper coatings.
(Jim Thorington)

GATX 85918
ICC103W
▲ Here is a car found at Columbus, Ohio in April 1962. It had been constructed that very month by General American. There is no known number series, but GATX 85937 and 85940 were similar.
(Paul C. Winters)

GATX 86127
DOT103W
◀ A photo taken in August 1971. Assembled by General American in November 1960 it was an 8,188 gallon tank car. It was probably one of 50 numbered as GATX 86110-86149.
(William Rosenberg)

GATX 86382
ICC103W

◀ A picture of a 10,080 gallon car taken in May 1971. It was first listed in the equipment register in October 1967 and became part of the set 86371-86404, though more cars may have been added.
(William Rosenberg)

GATX 86997
ICC103W

▲ Assembled by General American in May 1964, the car was caught on film that month. With a 9,969 gallon capacity, it was one of almost 100 cars placed within the 86900-86999 number set. Heyden Newport was a large chemical manufacturer acquired by Tenneco in March 1965.
(Paul C. Winters)

GATX 87781
ICC103W

▶ The car illustrated in this photo, which was taken in May 1971, would appear to be one of ten added in early 1963. The number series was likely 87778-87787. Its lessee was Hercules Incorporated (formerly the Hercules Powder Company, incorporated in 1912) a specialty chemical and explosives maker. *(William Rosenberg)*

GATX 96014
▼ A tank car caught on film in April 1969 at Birmingham, Alabama. At the time of its photograph, it was simply listed as one of 163 cars scattered somewhere within the GATX 95032-98498 number series. *(Jim Thorington)*

HCPX 1145
DOT105A500W
▼ In April 1965 Hooker Chemicals added this new reporting mark which was used for cars acquired through the Commonwealth Plan (mentioned on page 55). The marks stood for **H**ooker **C**ommonwealth **P**lan. Built for chlorine service in May 1966 the car was one of 50 numbered as 1104-1153. They were 10,600 gallon capacity cars. The picture was taken in Milwaukee, Wisconsin in February 1984.

(Don Reck, Gib Allbach collection)

HOKX 1352
ICC103W
▶ More Hooker cars but with different reporting marks. Hooker had used these reporting marks since the early 1920's (though as predecessor Hooker Electrochemical Company). The 10,100 gallon car shown here was first listed in the July 1964 equipment register. It was photographed in February 1973 and its series was 1340-1408.

(William Rosenberg)

HOKX 2030
DOT105A500W
▲ We're at Hearne, Texas on June 23, 1988. The chlorine tank car was constructed in mid-1963 and part of the 17-car set numbered as HOKX 2016-2032. *(Al Lanier, Jim Kinkaid collection)*

KPCX 3074 & 3123
ICC103W
▲◀ A couple of cars from the Koppers Company. At middle is one of 75 such cars in the 3000-3074 number group, first listed in 1951. This car was at Newark, New Jersey when it was photographed in August 1971. Of 10,000 gallon capacity, this particular car's home point was Warren, Ohio. In the lower position is a view of a 10,096 gallon car from the Koppers number group 3100-3309. In this view the photographer was at Elizabeth, New Jersey in September 1969. Note the external heater pipe connections on the tank heads of both of these cars.
(both, William Rosenberg)

75

KPCX 3148 & 3212
ICC103W

▲▶ Two more views of some of these cars, which were somewhat unique by having their placard boards offset from the tank shells. At top is one built in November 1951 by General American, and rated at 10,088 gallons. It was in Youngstown, Ohio on May 4, 1968. At middle is another one with an August 1951 manufacture date. It's photo was taken in November 1968 and was a 10,093 gallon capacity car.

(top, Dave McKay, Morning Sun Books collection; middle, Emery Gulash, Morning Sun Books collection)

KPCX 4117
DOT103W

▶ Here's a Koppers car from the 4100-4124 group. It's photo was taken in March 1972. These cars were constructed in late 1951 and were of nominal 12,500 gallon capacity. Koppers was a creosote and wood treatment supplier who derived its products from the distillation of coal tar. Most treated railroad ties, for example, come from this company.

(William Rosenberg)

LCIX 2565
DOT105A500W

◄ At Emporia, Kansas in December 1989. Operated by Liquid Carbonic the car was leased from Union Tank Car who constructed it in August 1988 under lot 6538. It was one of 25 in the set 2541-2565 assembled from March through August of that year. *(Jim Kinkaid)*

NATX 7172
ICC103A

◄ The photo was taken in Birmingham, Alabama in May 1958. This car was built by the Pressed Steel Car Company in July 1930 and is from the series 7100-7174. The company leasing the 10,050 gallon car was based in Louisiana and was formed in February 1940.
(Jim Thorington)

NATX 9300
ICC103W

▼ In January 1957 American Car and Foundry assembled this car for North American as part of the 6-car set NAHX 9300-9305. They were constructed under ACF lot 02-4847A. It was recorded on film in May 1966.
(Paul C. Winters)

NATX 9915
ICC111A100AW
◀ In January 1965 North American began listing the series 9900-9919, of which this car was a part of. Used for sulfuric acid service, it is thought that these 20 cars were acquired used. This photo was dated April 1969.
(Paul C. Winters)

NATX 12511
▶ In July 1964 the North American shop in Texankana, Texas assembled 15 cars of nominal 8,000 gallon capacity and numbered them 12500-12514. Here is one photographed on August 19, 1974. A close look will show that the ACI label is ready to fall off. Among other things, Pfizer's Chemical Division made products for the food industry. *(William Rosenberg)*

NATX 19538
ICC103
▼ It is believed to be one of 30 such cars assembled in late 1957 or early 1958 and numbered as NATX 19533-19562. At Waukegan, Illinois on May 31, 1965 the car, which was equipped with heater coils, was likely leased by Jefferson Lake Sulphur Company at one point in time (compare this car to NATX 7172 on the previous page).
(Owen Leander, Morning Sun Books collection)

NATX 19639
DOT103W
▼ This car was constructed in August 1960 and was placed on film in April 1975. Originally part of the 47 car set NATX 19600-19646, by January 1963 the series had been expanded to 97 cars and numbered as 19600-19696. Engelhard's Minerals and Chemical Division is largely devoted to kaolin production. *(Bernie Wooller)*

NATX 20320
ICC111A100W1
▲ At Birmingham, Alabama in June 1976. The tank was built by AMF in December 1965 and assembled by North American Car at Texarkana. As such it was one of 26 similar rubber lined cars in the series 20300-20325. *(Jim Thorington)*

NATX 21183
ICC111A100W1
◄ Photographed in August 1967 and of 20,539 gallon capacity, it was assembled in February 1962. There were ten cars in this group, which was numbered as NATX 21177-21186.
(William Rosenberg)

NATX 21521 & 21730
ICC111A100W1
▲▶ Two rather similar cars. The upper car, assembled in October 1963, was put on film at Emporia, Kansas in December 1989. The car was under lease to Westvaco and carrying something called "resonate". (This is likely a mis-stencil. Westvaco, renamed in 1969 from West Virginia Pulp & Paper, was primarily a paper and packaging company. The car is likely transporting "rosinate" used as paper sizing.) It was built by North American Car at Texarkana and was originally placed into the 15-car set 21517-21531. This was quickly reset by January 1964 as 21516-21530, apparently due to a numbering conflict. The 20,000 gallon car at middle was photographed in August 1967 hauling napthalene (an intermediate chemical utilized in the agricultural and textile trades). It was constructed by North American in February 1965 and was part of the 10-car set 21721-21730. Although quite similar to the one at top, note that this car's jacket is fully welded, while 21521 has a riveted flange running down the middle of the car.

(top, Jim Kinkaid;
middle, William Rosenberg)

NATX 27841
ICC105A200
▶ Trenton, New Jersey was where this tank car was spotted in November 1967 while on the Pennsylvania Railroad. It had been built in January of that year at North American Car in Texarkana. It would seem to have been one of 101 cars added to the existing NATX 27600-27793 number series (as NATX 27794-27984) in late 1966 and early 1967.

(William Rosenberg)

NATX 28064
DOT111A60ACW
▲ At Birmingham, Alabama in July 1974 is a brand new car, having been constructed by North American Car the previous month. Built at Texarkana, it was one of 20 cars within the 28048-28067 set. These were 23,800 gallon cars. *(Jim Thorington)*

PCIX 164
ICC105A500W
▼ In carbon dioxide service, it was photographed in August 1967. It was part of the PCIX 164-183 set of 20 cars added in July 1966 to an existing ACF-constructed fleet numbered as 150-163. Interestingly enough, by July 1968 this car (plus 6 others) went missing, perhaps due to wrecks.
(William Rosenberg)

PQX 857
AAR204
▲ Constructed by American Car and Foundry in late 1939 under lot 1954, it was one of ten cars numbered PQX 849-858. The cars were originally built as AAR103 cars, so obviously they had been converted at some point in time. Note the yellow journal box lids in this photo taken in November 1969! *(Paul C. Winters)*

RTCX 5294
ICC111A100W1

▶ The Republic Tank Car Company was the owner of this car. It was part of a 100-car set assembled by American Car and Foundry in November 1952 under lot 02-3952. As built, the cars (numbered as RTCX 5251-5350) were ICC105A300W specification cars. Over the years, Republic converted a number of these 11,000 gallon cars to other specifications, which is what happened to this car, photographed in January 1974. Rubicon (named after an Italian river) Chemicals, based in Louisiana, made chemicals used in everything from foam products to car bumpers. *(William Rosenberg)*

SHPX 364
ICC105A500W

▼ Used for chlorine service, it sits for a portrait in May 1965. This car was furnished by American Car and Foundry in March 1962 and was part of the 15-car set SHPX 354-368. *(Bernie Wooller)*

SHPX 718
ICC105A300W

◀ Another car in hydrocyanic acid service, with its attendant placard boards. The photographer caught this car in February 1964. American Car and Foundry assembled the car in May 1962. This may have been a single car order added to the previously listed 607-717 set.

(Paul C. Winters)

SHPX 3299

◄ American Car and Foundry assembled the car in November 1946 and it was used for chlorine service. Its history is unknown, but may have been part of the 4-car set 3295-3299. *(Bernie Wooller)*

SHPX 3600
ICC105A500W

▼ Coupled next door was another car under lease to Stauffer. American Car and Foundry assembled it in September 1947 and the car was also used for chlorine service. It was furnished under lot 3119, which was an order for 35 cars numbered SHPX 3600-3634 and originally constructed as ICC105A300W specification.
(Bernie Wooller)

SHPX 5790
ICC105A300W

◄ Obviously used in hydrocyanic acid service, this Shippers Car Line car was photographed in February 1964. It would appear that the tank was constructed in March 1959 by General American Transportation Company and was assembled as a complete railcar at ACF's Milton plant that same month. This was a single car order. *(Paul C. Winters)*

SHPX 6184
ICC103W

▲ Equipped with heater coils for caustic soda service, it sits at Marion, Ohio on July 30, 1967. American Car and Foundry's Milton, Pennsylvania plant built this car in March 1957 and it has a tank capacity of 8,069 gallons. It was one of 6 such cars assembled under lot 02-4867 and were numbered as SHPX 6180-6185.
(Dave McKay, Morning Sun Books collection)

SHPX 6724

◀ In chlorine service this car was photographed in August 1964, right after the ACF Milton plant's delivery. Between April 1964 and January 1965 ACF placed 143 similar cars in service and numbered them as SHPX 6654-6796.
(Calvin T. Banse, Morning Sun Books collection)

SHPX 12014
ICC111A60W1

▼ There were ten of these cars constructed by ACF in April and May 1962. They were numbered as 12010-12019 and were assembled under lot 72-5899. This 20,789 gallon car was found in May 1964 at Birmingham, Alabama. Chemstrand, a division of the Monsanto Company, is best known for its synthetic fiber and turf manufacturing. *(Jim Thorington)*

SHPX 14277
ICC103W
▶ One month later, in June 1964 and also at Birmingham, Alabama. As may be seen, it was constructed in December 1959 by American Car and Foundry and was a 10,119 gallon car. First listed in April 1959, the 33-car group was marked as SHPX 14260-14292. Named after founder Henry Reichhold, the company stenciled on this car is a maker of composites and coatings.
(Jim Thorington)

SHPX 15686
ICC103W
▼ Here's a photo taken in Columbus, Ohio in April 1962. The car was in the Pennsylvania Railroad's Grandview Yard and was constructed in July 1955 by American Car and Foundry under lot 02-4476. The 5 cars in this group were listed as SHPX 15684-15688.
(Paul C. Winters)

SHPX 85184
ICC105A500W
▲ In liquid carbon dioxide service, the car was photographed in May 1968. It was constructed in September 1967 at American Car and Foundry's Milton plant under lot 71-13342. This was a two car order for 18,600 gallon cars, the other being 85183. *(Paul C. Winters)*

SHPX 90244
ICC111A100W1

▼ Although the car shows markings stating that it was constructed in March 1947 by ACF at Milton, Pennsylvania, the car's number is not original to it. When photographed in January 1969 it was one of ten cars apparently converted from the original ICC105A300W specification. This occurred around October 1964 when these renumbered cars were first listed in the equipment register. The new series was SHPX 90236-90245. *(Paul C. Winters)*

UTLX 27499
DOT105A500W

▲ At Emporia, Kansas in December 1989. It belonged to a set numbered as UTLX 27458-27507. Assembled by Union Tank in East Chicago in April through June 1984 the cars were rated at 20,030 gallons. This set would eventually be moved over to LCIX 2583-2607 markings. *(Jim Kinkaid)*

UTLX 43206

▶ This tank car was constructed in October 1957, perhaps by General American. The photo was taken in April 1972 and information suggests that it was part of a 14-car set numbered as UTLX 43200-43213. They were 11,000 gallon cars. Note that the car is missing part of its specification (which, based on the commodity, was likely DOT111A100W3).

(William Rosenberg)

UTLX 82099
◀ A tank car constructed in December 1960 who's photograph was taken on November 21, 1962 at Columbus, Ohio. Rated at 16,500 gallons, it was one of three numbered as 82097-82099. *(Paul C. Winters)*

UTLX 84287
DOT105A300W
▼ At Birmingham, Alabama in December 1971. It is carrying liquid nitrogen fertilizer solution and was assembled in March 1959. Farmers Chemical Association was a producer of ammonia products and was based out of Chattanooga (note the incorrect spelling stenciled on the car).
(Jim Thorington)

UTLX 86637
DOT103ALW
▼ When photographed in December 1985 the car was assigned to Domino Sugar service. It was assembled in January 1956 with an aluminum tank made by Graver Tank and Fabricating, who supplied tanks for Union Tank Car for some years. *(Jim Rogers)*

UTLX 87479
ICC105A100ALW
▲ On February 25, 1962 this car was caught on film at Columbus, Ohio while in nitrogen fertilizer service. It was constructed in September 1956. *(Paul C. Winters)*

UTLX 87628
▲ A car that carried 10,654 gallons of hydrocyanic acid. It was built in mid-1957.
(Al Lanier, Jim Kinkaid collection)

UTLX 87772
DOT105A500W
◄ Used in chlorine service, it was photographed in June 1971. Built in December 1962 the car carried 10,669 gallons. It might have been a part of the UTLX 87750-87859 number set. *(William Rosenberg)*

UTLX 87828 & 87837
DOT105A500W

▶▼ A couple more GAF paint schemes. The upper photo was taken in October 1967 while the middle one was caught on film in January 1968. While there is no history available on the upper car, UTLX 87837's tank had been furnished by Graver Tank and Manufacturing in January 1963 and assembled by Union Tank Car that April.
(both, William Rosenberg)

VMCX 5501
DOT105A500W

▲ A Vulcan Chemicals car in chlorine service at Wichita, Kansas on September 13, 1992, which was its home point. It was fabricated by ACF's Milton plant in June 1966 as part of the VMCX 5501-5525 group. The reporting marks stand for "Vulcan Materials Company." *(Jim Kinkaid)*

VMCX 5514
DOT105A500W
▼ Another in the earlier gold and blue scheme. This one sat at the Vulcan Chemical plant outside of Wichita on March 30, 1991. Cars with VMCX reporting marks were actually owned by the "Commonwealth Plan," (mentioned on page 55). *(Jim Kinkaid)*

WCHX 10045
DOT111A100W1
◀ It Baltimore, Maryland in February 1988 the tank car was constructed in January 1955 and was of 10,921 gallon capacity. The Walter Haffner Company, who owned these reporting marks, first began listing the car in July 1973 as one of ten in the group 10041-10050, so it had to have been acquired used. *(Jim Rogers)*

WCIX 102
ICC103W
▼ Photographed in April 1969, this was a single car listed under Witco Chemical's Equipment Register listing. Although there is no new date stenciled on the car, it was first reported in January 1961 and was a 10,000 gallon car. Its commodity, liquid phthalic anhydride, is a constituent of PVC and fiberglass manufacturing. *(Paul C. Winters)*

Small Cars

This section is devoted to single compartment tank cars of approximately 4,000 gallons or less with full platforms.

ACFX 813
ICC103
▼ Constructed in September 1960 by ACF, it sits in Council Bluffs, Iowa on August 21, 1987. This car has a tank capacity of 4,049 gallons and would appear to be one of six similar cars built as ACFX 808-813. *(Jerry Bosanek, Jim Kinkaid collection)*

ACFX 815
DOT103W
▲ A rather plain car sits in the Santa Fe yard at Wichita, Kansas in March 1989. It was constructed in August 1950, and was likely one of two listed as cars 814-815. The car shown here has a tank capacity of 4,065 gallons. *(Jim Kinkaid)*

ACFX 888
ICC103
◀ A similar car found in Birmingham, Alabama. Built in January 1959 it was shown as part of the group 885-860 in the equipment register. As can be seen, it has a tank capacity of 4,071 gallons. *(Jim Thorington)*

ACIX 1005
DOT105A300W
▲ Owned by Arkansas Chemicals Incorporated and carrying bromine, it was put on film in June 1984. First listed in January 1962, it was probably constructed by General American in November of the previous year. A 4,400 gallon capacity car, this was the only such car reported in the equipment register. *(Al Lanier, Jim Kinkaid collection)*

BECX 96 & 495
ICC103ALW
▲▶ At middle is a hydrogen peroxide car with an aluminum tank found in Muhlenberg Township, Pennsylvania on March 5, 1977. Constructed in July 1952 by American Car and Foundry, it has a tank capacity of 4,027 gallons. Furnished under builder's lot 02-3846, it was one of five cars sold as BECX 93-96 and BECX 99. Meanwhile, at bottom is another ACF-supplied car assembled in March 1957 under lot 02-4918 as part of the set 494-497. It sits at Grand Junction, Colorado on May 25, 1973. The reporting marks indicate the original company name, Buffalo Electro-Chemical Company, Inc., which by 1957 became Becco Peroxide, part of the FMC corporation.
(middle, Craig Bossler; bottom, Jim Kinkaid collection)

DOCX 3003
DOT105A300W

◀ A 3,000 gallon car with DuPont of Canada reporting marks. This little car was found at Winnipeg in July 1993 and was constructed by General American in March 1949. Because the reporting marks did not exist until the 1960's, the car (part of the DOCX group 3001-3005) was acquired used. *(Jim Kinkaid collection)*

DOWX 52302
ICC105A300W

▲◀ Two views of the same car, but wildly different paint schemes. At middle is the Dow Chemical Company car at Birmingham, Alabama in November 1966 while the bottom photo was view taken on April 9, 1972 at Thorndale, Pennsylvania. The tank on this car is only of 2,255 gallon capacity: they just don't come much smaller! It was one of two cars ordered five months apart but constructed at the same time (as DOWX 52301-52302) in January 1949 and nickel lined for bromine service.

(middle, Jim Thorington; bottom, Craig Bossler)

DUPX 6420 & 6426
DOT105A300W

▼▼ At top is a DuPont car found sitting at Reading, Pennsylvania on January 25, 1989. The photographer notes that its tank capacity is 3,021 gallons and the car shows an August 1956 manufacture date. At middle is a car constructed in January 1957 by General American Transportation Company and carries 3,018 gallons. Based on equipment register listings, the two cars would appear to be part of the 15-car set 6413-6427.
(top, Craig Bossler; middle, Emery Gulash, Morning Sun Books collection)

DUPX 7411

▶ Between April 1955 and October 1956 the DuPont company acquired 22 of these cars and listed them as DUPX 7401-7422. The car in this photograph would appear to have been assembled in May 1956. Due to the time frame involved, it may be that this set is comprised of various smaller orders grouped together. This photo was taken in April 1970.
(Jim Kinkaid collection)

EBAX 3163
ICC105A300W
▲ A photo taken in July 1971. It shows a 3,000 gallon car supplied by General American in June 1960. It would seem that it was part of a 46-car set added at that time as EBAX 3140-3185 (added to the existing 3029-3139 series). Of note is the fact that one of the smaller tank cars on the road is coupled to one of the largest! *(William Rosenberg)*

EBAX 4404
DOT105A300W
▲ Unusual for such a small tank is the full length walkway. It was photographed at McNeil, Arkansas on March 27, 1984 and was a brand new car, having been constructed the previous December by General American. This was one of two in the set 4403-4404.
(Al Lanier, Jim Kinkaid collection)

GATX 1102
DOT105A300W
◀ A 2,249 gallon tank car sits at El Dorado, Arkansas on February 15, 1983. It was a single car first mentioned in the July 1965 issue of the *Official Railway Equipment Register* where it was noted that it was in bromine service.
(Al Lanier, Jim Kinkaid collection)

GATX 39002
DOT105A300W
▲ Another car at El Dorado, Arkansas, though on March 27, 1984. Also in bromine service, it had been constructed by General American in December 1964 as one of two in the set 39002-39003. Of note is the circular walkway surrounding the manway.

(Al Lanier, Jim Kinkaid collection)

GATX 60144
ICC103W
◀ A 4,043 gallon car that was constructed by General American in November 1954 is shown here. From the available register information, it may have been one of 11 built as lined cars and marked with various car numbers.

(Emery Gulash, Morning Sun Books collection)

GATX 73721
DOT103A-ALW
▼ A car with an aluminum tank sits at Pine Bluff, Arkansas on December 1, 1983. Used in hydrogen peroxide service, it was likely one of seven such cars built in early 1983 and numbered as GATX 73715-73721.

(Al Lanier, Jim Kinkaid collection)

GATX 87013
DOT103ALW

▶ A photograph taken at Muhlenberg Township in Pennsylvania on March 5, 1977 of a tank car built in September 1957 and in hydrogen peroxide service. The car belongs to the GATX series 87000-87022. These 23 cars appeared throughout late 1957 and all 1958. *(Craig Bossler)*

GLKX 6009
DOT105A300W

▼ El Dorado, Arkansas was the location of this car when this Great Lakes Chemical-owned car was caught on film on March 27, 1984. It was one of eight in the 6006-6013 set, all acquired second-hand and in bromine service. While the paint job suggests a complete repaint there in June 1978, that is impossible since this reporting mark did not appear until the April 1982 equipment register. The paint shop simply carried over the prior reweigh date.
(Al Lanier, Jim Kinkaid collection)

HMHX 1976
ICC103

◀ Few railcars were painted in American bicentennial colors for the 1976 celebration. Even fewer tank cars were so done, making this car (owned by the Tank Car Corporation of America) rare. Photographed in July 1976, it was apparently renumbered from HMHX 4000, which had been acquired used in October 1968 (since it had been constructed by ACF in March 1942). For some reason the equipment register never did note the car number change of this car. Its origin remains unknown.

(Jim Kinkaid collection)

MONX 404
DOT103A

◀ Such a small car (at 4,050 gallon capacity) yet such a large dome platform: it is so wide that the ladder tilts outward to get to it! Owned by the Monsanto Company it was at Reading, Pennsylvania on July 23, 1974 and was constructed in February 1935. Originally tin lined, it may have been a single car order.

(Craig Bossler)

SACX 500
ICC105A500W

◀ Assembled in October 1936 and photographed in May 1969, it is part of the 500-509 number set. This series was originally built as CACX 500-509 for Columbia Chemicals, with them being remarked SACX (for Columbia Southern Corporation) in January 1954. Some cars were eventually remarked PPGX (for PPG Industries) and continued in service into the mid-1970's.

(William Rosenberg)

SHPX 474

▼ Photographed in August 1963. It's hard to say with any certainty, but the car might be one of three numbered SHPX 472-474 and possibly constructed shortly before the photo was taken. *(Paul C. Winters)*

TRMX 521
◀ A car owned by The Railroad Museum at Telford, Pennsylvania. With its circular manway platform, the car awaits scrapping at Reading, Pennsylvania on November 4, 1992. It was originally constructed by General American in February 1965 but first listed under TRMX markings in April 1977 as a Kanigen (a trademarked name for a form of nickel plating) lined car. *(Craig Bossler)*

TRMX 6502
DOT103W
▼ Another Railroad Museum car, found at Reading, Pennsylvania on November 24, 1992. It has a 4,026 gallon tank and was fabricated in April 1957 by American Car and Foundry, probably under lot 02-4848 as NATX 6502. Its photo was taken on November 24, 1992 and first showed up in the April 1975 equipment register. *(Craig Bossler)*

VELX 6010
DOT105A300W
▼ Our last small tank car is this Velsicol Chemical Corporation 2,420 gallon insulated car found at Memphis, Tennessee on February 11, 1981 for bromine service. It was one of four acquired second hand from prior owner Michigan Chemical. As the car has a November 1960 build date stenciled, its origin is with Michigan Chemical's MCTX 6007-6013 group. *(Al Lanier, Jim Kinkaid collection)*

6-Axle Cars

CELX 6414
DOT111A60ALW1
▲ Celtran is the only owner of 6-axle tank cars with full underframes in our volume. The reasoning is due to the fact that these cars were constructed with aluminum tanks. This example, which carried 35,438 gallons of acetic acid, was at Slidell, Louisiana on February 26, 1994. Part of the CELX 6400-6458 series, it was a non-insulated car equipped with heater coils manufactured by General American at Sharon, Pennsylvania. Production spanned from June through December of 1968.
(Lee Yoder)

CELX 10424 & 10425
DOT111A60ALW1
▶▼ Another group of uninsulated Celtran 6-axle tank cars are these ones assembled by AMF Beaird in Shreveport, Louisiana from October 1969 through February 1970. The two seen here were at Birmingham, Alabama in May 1971. Also fitted with coils and used in acetic acid (and acetic anhydride) service, they were part of the 10,400 gallon group numbered as 10400-10438. *(both, Jim Thorington)*

Multi-Compartment Cars

ACDX 8903
ICC103

◄ South Modena, Pennsylvania was the location of this car when it was put on film on September 8, 1974. It was most likely there to be scrapped. As may be seen, it was built by ACF in March 1935 (under lot 1403) and it had heater coils fitted. At that time it was built as BMX 903 and sold to the Barrett Division of Allied Chemical. This 6,000 gallon insulated car would later be remarked as PLCX 903 and eventually restenciled as shown.
(Craig Bossler)

ACDX 8923
ICC103

► Another ex-BMX car found in South Modena, Pennsylvania. It was found there on September 3, 1974, again likely for scrapping. Also built by ACF in March 1935, it was assembled under lot 1402. Like 8903 above, it too went to PLCX markings (as car 923) prior to the final ACDX restenciling. It is a non-insulated car fitted with heater coils.
(Craig Bossler)

ACFX 12
DOT103W

▼ At Council Bluffs, Iowa on September 7, 1987. It was fabricated in February 1959 and was assigned to the Pennzoil Company. The insulated car would appear to have been a single car order. The car does have heater coils.
(Jerry Bosanek, Jim Kinkaid collection)

101

CPVX 101
ARA III

▼ Found in a museum at Belton, Missouri on March 8, 1992, it carries the paint scheme used in many years in service. It was an uninsulated, coiled car built in January 1917. Apparently it was acquired used by Cook Paint and Varnish, the owner, since it was not listed in the equipment register until the 1930's. Note the cover plate where a third dome would reside: apparently this has always been a twin compartment car (note how that plate spans the area where the two internal compartment heads are riveted). *(Jim Kinkaid)*

DUPX 911

▶ A "Community Awareness Emergency Response" training car seen on the move at Rochester, Pennsylvania in June 1990. The car first appeared in the October 1988 edition of the *Official Railway Equipment Register* but its heritage is unknown. It would be eventually be given the AAR class "MT" assigned to training cars. *(Jim Kinkaid)*

DUPX 14004
AAR211A100W1

▲ What might look like a five compartment car was really a single compartment car with four additional manways for additional access to the tank. The clue is the diamond symbol at the center manway showing it to be the approved manway along with the single discharge pipe. Found in Memphis, Tennessee on October 3, 1983 the car was part of a four car set numbered 14003-14006. These 10,400 gallon cars first appeared in the July 1965 register. *(Al Lanier, Jim Kinkaid collection)*

GATX 844
DOT103W
▲ Here's an insulated car found in Wichita, Kansas in August 1989. It was under lease to the Lubrizoil Corporation and is thought to have been assembled in September 1963 as a single car order. It was fitted with coils. *(Jim Kinkaid)*

GATX 849
ICC103W
▼ An uninsulated car found in Decatur, Alabama in May 1966. There is no information on the car available, but from left to right the tanks held 2,046, 2,013 and 2,040 gallons respectively. *(Bernie Wooller)*

GATX 936
◀ An unusual example found in Marion, Ohio on June 22, 1962. Commonly called wine cars by modelers, this one may have been in some other service since, while it was insulated, it also was fitted with heater coils. The first mention of the car was circa October 1950 as part of the 912-943 car set.
(Paul C. Winters)

GATX 1058
ICC103
▲ Photographed in May 1974, the Dow Company logo is barely visible up under all the walkway grating shadows. Apparently constructed in the early 1930's, it would seem to be part of the 45-car set GATX 1032-1076. It is non-insulated and has heater coils.

(Rail Data Service, Jim Kinkaid collection)

GATX 1406
ICC103
▲ Obviously having problems with the center dome, this car was caught on film in October 1970. It was built by General American in May 1935. Available information suggests that it may have been part of the 1401-1412 set of cars. *(Rail Data Service, Jim Kinkaid collection)*

GATX 1643
ICC103
▶ On May 11, 1962 the car was found at Columbus, Ohio. It was built in August 1930 by General American Transportation Company and under lease to a company that was a leading producer of high fructose corn syrup.

(Paul C. Winters)

GATX 3040
ICC103W
▲ Insulated and fitted with heater coils, it sits at South Edmonton, Alberta in Canada on June 23, 1982. It was first listed in the register in July 1973 as a possible single car order.
(Matt Herson, Jim Kinkaid collection)

GATX 25465
◀ A photograph taken of a non-insulated car found at Belt Line Junction in Pennsylvania on September 14, 1990. The asymmetric tank size is unusual.
(Craig Bossler)

GATX 25927
ICC103W
▼ At Birmingham, Alabama in June 1971, it was equipped with heater coils and was non-insulated.
(Jim Thorington)

GATX 31211
DOT103W
▼ Although there is no manufacture date in view, the car first appeared in the April 1970 issue of the equipment register. Non-insulated, it might have been one of two such cars (the other being GATX 31208). This car was at Birmingham, Alabama in April 1980. *(Jim Thorington)*

GATX 31205
DOT103W
▶ We are at South Bend, Indiana on May 16, 1984. The car was a part of the 31200-31214 set, which were first reported in the equipment register in April 1960. Non-insulated and with coils, note that it still has its friction bearing trucks at such a late date. *(Lee Yoder)*

GATX 35625
DOT111A60ALW1
▼ This illustration was taken in the Santa Fe yard in north Wichita, Kansas in May 1988. An insulated car equipped with heater coils it was part of the four car GATX set 35625-35628 constructed in February 1971. *(Jim Kinkaid)*

GATX 38832
DOT103W
▼ Much like DUPX 14004 shown earlier, this is actually a single compartment car designed with four additional manways for interior access. It was at Hearne, Texas on June 23, 1988 and carrying chromic acid, a substance used primarily for chrome plating.
(Al Lanier, Jim Kinkaid collection)

GATX 39918
DOT103W
▼ A single car order built to an unusual design. The car was at Muskogee, Oklahoma in July 1988 and had been assembled in April 1965. The heater coil caps can be seen dangling underneath this insulated car. *(Jim Kinkaid)*

GATX 52341
DOT103W
▲ Here's a car found sitting at Hearne, Texas on March 12, 1988. *(Al Lanier, Jim Kinkaid collection)*

GATX 65133
DOT103W
▶ In October 1963 the equipment register first reported this car as a single car listing. The non-insulated car, fitted with heater coils, was at a place called Laurel siding in Pennsylvania on December 24, 1982.
(Craig Bossler)

GATX 70490
DOT103W
◀ Wyomissing, Pennsylvania was where this insulated car was found when it was photographed on May 7, 1972. The car also has heater coils. Markings indicate that it had been assembled by American Car and Foundry in August 1941, probably for someone other than General American originally. At the time of the photo it was one of five in the set 70487-70491.
(Craig Bossler)

GATX 70501
AAR203W
▲ Bethlehem, Pennsylvania was the location of this insulated car when photographed on December 25, 1971. It was constructed by General American in late 1950. Interestingly, like the car above, the placard board above the right hand truck gives information on what commodity is in each compartment. *(Craig Bossler)*

GATX 70503
◄ Although at first glance this car looks similar to GATX 70501 opposite, there are differences. While the car shows markings from General American in June 1950, that may be a possible conversion date from a prior specification. The photograph was taken in April 1969 and its possible that it was part of a two car set (70504 being the other).
(Paul C. Winters)

GATX 70532
ICC103W
▼ An insulated and coiled car at Birmingham, Alabama in March 1969. Of note is how the tank compartments were marked.
(Jim Thorington)

GATX 70598
ICC103
▼ Not much is known about this car except that the photograph was taken in August 1971 while in Gliddon paint service. At the time of its photo, the insulated car (fitted with coils too) was somewhere within the 70598-70696 number group. *(Paul C. Winters)*

GRYX 168
ICC103
▼ A John H. Grace Company car found in Cleveland, Ohio on March 17, 1974. As can be seen, the uninsulated car was constructed in September 1941. It was put into the number series 100-199 and had heater coils. Since this number series had existed at least as far back as 1932 and Grace didn't bother to put car quantities in the equipment register, it is impossible to determine when the car was added or even how many were in existence. *(Dave McKay, Morning Sun Books collection)*

GRYX 832
ARAIII
▶ A non-insulated car whose photo was taken in August 1967. It would appear to have been constructed in August 1927 and was shown as a two car addition to the equipment register in October 1930 as GRYX 831-832 (although due to reported gallonage issues, this car may be a replacement for the original 832).
(William Rosenberg)

HHCX 5016
ARAIII
▼ At Birmingham, Alabama in May 1970 is a car owned by the Champlin Petroleum Company. A 1927 build date can barely be made out under all that grime. It's possible that it was part of the 5011-5017 number series added circa late 1927 or early 1928. The reporting marks, by the way, refer to founder H.H. Champlin, of Enid, Oklahoma.
(Jim Thorington)

HMHX 3800, 4201 & 6013
▲◄▼ Some non-insulated cars operated by the Tank Car Corporation of American found at Modena, Pennsylvania for scrapping. At top is one there on March 31, 1973. While there is no build date evident, the 8,000 gallon car had been listed in the equipment register at least as far back as mid-1938 as part of the three car set 3800-3802. Car 4201 was there on August 11, 1974. The additional straps laid over the top of the tank is noteworthy as is the rust from tank straps moving around over the years. The car, with ICC103 specification markings, was acquired used. It was one of two (4200-4201) that first appeared in the July 1962 edition of the *Official Railway Equipment Register* as 4,500 gallon cars. And at bottom is a car that was one of two added in April 1956. The two cars, numbers 6011 and 6013, were added to the already existing HMHX 6012. Note that this 6,000 gallon car still has its wooden blocks supporting the tanks when its photo was taken on April 4, 1973. The height of the domes is unusual.

(all, Craig Bossler)

NATX 3448
ARAIII
▲ A North American Tank Car Company car at Marion, Ohio on March 17, 1963. It shows a manufacture date of April 1920 by Standard Tank Car. By the time of this photo it was the only such car reported in the register and was a non-insulated car. The heater coil caps can be seen hanging from the car. The company leasing the car was a manufacturer of sealants and coatings. *(Paul C. Winters)*

NATX 16017
▲ Built in November 1960 it has a sparger system installed (which injects a gas into the product to help unload it). Like several other cars in the section, it is a "pseudo" multi-compartment car. The uninsulated car really has but one tank, with additional access fittings also installed. May 1976. *(Paul C. Winters)*

NATX 18600
DOT103W
▶ Under lease to Pennzoil is a car found in July 1989 and constructed in December 1959. The non-insulated car was part of the two car NATX set 18600-18601 and had a total capacity on 6,094 gallons. It was equipped with heater coils.
(Rail Data Service, Jim Kinkaid collection)

NATX 18801
DOT103W

◀ The photographer found this non-insulated one at Mansfield, Ohio in April 1979. It was also from a two car order (resulting in NATX 18800-18801). Constructed in November 1959 the tank had been supplied by the J.B. Beaird Co. (a vessel manufacturer in Shreveport, Louisiana) while North American supplied the frame and assembled the eventual car.

*(Emery Gulash,
Morning Sun Books collection)*

NATX 18850
ICC103W

◀ A single car order found in July 1968 and constructed in December 1959. Like the car above, its tank was built by the J.B. Beaird Co. and the car finished by North American. It is insulated and has heater coils.

(Jim Thorington)

NATX 18902
DOT103W

▼ Yet another single car order resulted in this car, which was uninsulated and had heater coils fitted. Photographed in May 1988 it was built in February 1960.

*(Rail Data Service,
Jim Kinkaid collection)*

NATX 23092
ICC111A100W1

▼ A car caught on film in February 1969 at Birmingham, Alabama. It was part of the NATX set 23091-23093. These were non-insulated cars fitted with heater coils. *(Jim Thorington)*

NATX 24002
ICC111A100W1

▼ Another car with unusual partitioning. Found in Reading, Pennsylvania in May 1970, it was constructed in October 1960 by North American at Texarkana, Texas. The left hand compartment holds 5,217 gallons while the larger one to the right holds 15,482 gallons: the weld seams adjacent to the ladder indicate where the internal tank heads lie. The Publicker Chemical Company leased this car, which was a one car-only design. *(Jim Thorington)*

NATX 24215
DOT111A100W1

▼ A non-insulated car moving through Kansas City, Missouri on March 26, 1992. Equipped with coils it appears to have been built in March 1965. If interested, the left hand tank holds 8,236 gallons and the right hand one holds 12,320. Similar cars were assembled throughout early 1965 resulting in the NATX number series' 24207-24213 and 24215-24220. *(Jim Kinkaid)*

NATX 25801
DOT103W
▲ At Wichita, Kansas in July 1988. Assembled in November 1965 at Texarkana, Texas, it was one of only two such cars built. The other car was NATX 25802: both were insulated and had heater coils installed. *(Jim Kinkaid)*

NWAX 1006
ICC111A100W1
▶ Its obvious what the reporting marks stand for! Found in December 1975 and constructed by General American, probably in October 1963, it was part of the three car series 1004-1006 of insulated cars. While they cannot be seen here, given the commodity, the cars would have had to have heater coils installed. *(Rail Data Service, Jim Kinkaid collection)*

QSMX 180
ICC111A100W1
▼ Likewise, "QSM" stands for Quaker State Motor Oil. An example resides at Thorndale, Pennsylvania on May 12, 1973. It was constructed by General American in August 1963 as a single car order. The car was uninsulated and did have coils applied. *(Craig Bossler)*

RSPX 837
▲ In January 1969 the car was in Tullahoma, Tennessee. While the 8,000 gallon insulated car appears to have markings indicating a May 1937 manufacture date from American Car and Foundry, it was not reported as being operated by owner RSP Car Company until July 1966. At that point in time it was one of seven cars in the set 832-838. *(Lee Yoder)*

SHPX 4402
ICC103
◀ Also at Tullahoma was this Shippers Car Line car, found on April 7, 1974. It would seem to have been a single car order built by ACF under lot 1877 in 1939. The car carried a total of 4,000 gallons. *(Lee Yoder)*

SHPX 4408
ICC103
▲ Another example found in April 1970. This one was part of the five car set SHPX 4406-4410, all constructed under ACF's lot 2142 around late 1940. The company Detrex, in business since 1920, is a producer of cleaners and solvents. *(Jim Kinkaid collection)*

SHPX 6313
ICC103
▼ Constructed by ACF in April 1937 under lot 1659 the car was photographed in January 1955. It was from the SHPX series 6301-6340, which were 6,000 gallon insulated cars. The heater pipe caps are quite obvious in this view. *(Emery Gulash, Morning Sun Books collection)*

UTLX 82
ICC103
◀ Another ACF-built car, but fabricated in May 1934 and uninsulated. It was at Pottstown, Pennsylvania on March 25, 1975. This car is not quite symmetrical; the far tank holds 2,539 gallons while the near one holds 2,044. While marked as having been supplied by American Car and Foundry, it is not in the builder's records.
(Craig Bossler)

UTLX 900
ICC111A100W1
◀ It was found in September 1964 and carries 5,139 gallons of chocolate syrup in the "A" compartment and 5,183 gallons in the "B" one. This was a single car order manufactured in June 1961. Insulated, the car almost surely had heater coils fitted. *(William Rosenberg)*

UTLX 1340, 3591 & 3756
▲▲▲ Some examples of run-of-the-mill three compartment, non-insulated Union Tank cars. At top is one example found at Decatur, Alabama on December 7, 1974. At middle is another car photographed at Edmonton, Alberta in Canada on June 23, 1982. Markings indicate that it was built in 1941. The firm leasing this car is in itself a major Canadian railcar lessor. Our last multi-compartment car was constructed in September 1936. It was caught on film at Rockville Bridge, Pennsylvania on September 6, 1976. All of these cars were simply lumped into the UTLX 1-9999 series, showing that it is nearly impossible to utilize Union Tank car listings for historical research purposes!
(top, Lee Yoder; middle, Matt Herson, Jim Kinkaid collection; bottom, Craig Bossler)

The "TMU" Cars

AAR class TMU is assigned to cars that originally had multiple removable tanks mounted on flat cars, typically used for chlorine. Eventually the class was also assigned to the helium cars, which have multiple tanks that are not removable. Although the flat car design seems antiquated, having been around since the 1930's (at least), surprisingly enough Pennwalt carried three on the books until as late as January 1988.

MHAX helium cars
DOT107A

▲▲▲◄► Here are four photos of various helium cars used by the United States Bureau of Mines to move compressed helium around the country. At top is one with original friction bearing trucks found in Tullahoma, Tennessee in June 1963. It is marked as being built in July 1955 and was part of the MHAX 1004-1078 group (this particular car was probably one of 17 supplied by ACF under lot 21-4481 as replacement cars). Meanwhile, we see the top of car 1171 while underway at Kansas City, Missouri on March 26, 1992. This example was built by Magor Car (builder's lot W-5200) in September 1961 as part of the series 1164-1173. Detail views of each end are also illustrated. The "B" end of MHAX 1121 (from set 1119-1138, built by ACF in February 1959) was taken at Wellington, Kansas in April 1990 while the opposite end (of MHAX 1188) was snapped at Huntsville, Alabama in May 1970.

(top, Lee Yoder; middle and "B" end, Jim Kinkaid; "A" end Bernie Wooller)

MHAX 1180
DOT107A

▼ A nice shot taken from ground level, showing the heavy duty trucks used on these cars due to their high weight. Contrary to urban legend, they do *not* get lighter as they are loaded! Our example was found scooting along at Tullahoma, Tennessee on June 1, 1974. The car can haul 318,000 cu.ft. of compressed helium. It was assembled as part of the set 1174-1196, supplied by American Car and Foundry in October and November 1961 under lot 61-5718. *(Lee Yoder)*

ACFX 426
ICC106A500

▼ Assembled by American Car and Foundry in April 1957 the car was part of the 11-car group originally marked as SHPX 423-433, assembled under lot 02-4872. It sits in Wyomissing, Pennsylvania on March 25, 1979 while under lease to Chemetron. *(Craig Bossler)*

ACFX 522
ICC106A500

▶ Another example found at Belt Line Junction on October 3, 1976. The car first appeared as part of a 23-car series SHPX 504-526 in late 1954. An interesting fact about this type of car: the specification is not marked on the car itself, but rather stamped on each cylinder.
(Craig Bossler)

JCIX 131
ICC106A500

◀ Owned by Jones Chemicals Incorporated, on June 7, 1976 it was caught at Lyons, Pennsylvania. The cars go back as far as March 1937 when Columbia Alkali took delivery of CACX 200-223. These were later remarked as PPGX (keeping those numbers, to which more cars were added) and once again as SACX 115-141 (SACX being the reporting mark for Southern Alkali, another constituent of Pittsburgh Plate Glass). Eventually, in July 1976, Jones Chemicals received 13 of these cars with one car remaining well into the early 1980's. *(Craig Bossler)*

RTCX 340
ICC106A500

▼ The January 1954 equipment register reported that Republic Tank Car acquired 43 of these cars from an unknown source. The car numbers ranged from 308 up to 374, with many car numbers not used. Here is one found in November 1964. *(Bernie Wooller)*

SHPX 216
ICC106A500

▼ American Car and Foundry assembled this car as part of the 10-car set SHPX 208-217 under 3173 around late 1946 or early 1947. It was seen in 1970. *(Rail Data Service, Jim Kinkaid collection)*

The "TW" Wooden Cars

These are often called pickle cars by the hobby industry. They do indeed have specifications and surprisingly enough, two wooden tank cars remained listed in the *Official Railway Equipment Register* until April 1980.

NFPX 22
ICC108
▶ The National Fruit Product Company owned this car. It was first reported in the July 1939 equipment register along with mate NFPX 20. The car was in use at least through late 1970 when it was donated to the Strasburg (Virginia) Railway Museum. In this photo at that place, taken in November 1986, it would appear to be in its last revenue paint scheme, though the "keep off" notations were likely applied by the museum. *(Jim Rogers)*

SBIX 1678
ICC108
▶ Operated by Standard Brands, Incorporated, this car was photographed in April 1964, still very much in revenue service The tank has an April 1958 construction date (which is probably the date that this car was rebuilt into a wooden tank car) while the underframe shows an October 1925 manufacture date. It was first mentioned in the July 1958 register. *(Paul C. Winters)*

The "XT" Box Tank Cars

Although these cars are built around house car designs (either ordinary-looking box cars or perhaps express car designs) when in ordinary freight car service they are AAR classified "XT" and considered tank cars. One unusual item here is that while some carry pressure vessels to hold liquefied gasses, there are no specifications stenciled on the outside of these cars, as the photos will illustrate.

AROX 123
AAR204
▶ Airco (previously named the Air Reduction Company) 123 was found on the Louisville and Nashville Railroad in July 1974 at Huntsville, Alabama. While it has 3,376 cu.ft. capacity markings, the interior was filled with cryogenic tanks. These cars were leased by General American and were in the AROX series 100-199.
(Bernie Wooller)

AROX 136
AAR204
▲ An earlier paint scheme, placed on film in July 1966. General American built the cars under Build Order 8109 in March 1957.
(Rail Data Service, Jim Kinkaid collection)

CMWX 163
◄ AAR class "BMT" was for bulk milk cars used in express service. However, the car illustrated here no longer carried milk and was apparently used in freight service, so it is included as an "XT" car. Photographed in October 1967 it was equipped with marker brackets at the corners and buffers for passenger train use, though the steam and signal lines would appear to have been removed. There were 16 of these cars first reported in the April 1960 equipment register, listed as car numbers 150-165. They were described as 6,000 gallon capacity cars.
(Emery Gulash, Morning Sun Books collection)

GPEX 803
▼ Another ex-milk tank is shown at Council Bluffs, Iowa on February 5, 1955. Besides the high speed trucks and marker brackets, it also has evidence of end sill buffers (apparently removed, along with the steam and signal lines). General American-Pfaudler listed this car as part of the 801-804 group, rated at 7,600 gallons. *(Lou Schmitz)*

▲ A view taken at the Arnold Engineering Development Center's Air Force Aerospace Test center, near Tullahoma, Tennessee, in March 1968. Leading the way is a Baldwin VO1000 switcher (later retired and sent to the Tennessee Valley Railroad Museum). This string was delivering a large load of liquid nitrogen, although rocket fuel was also delivered there. *(Lee Yoder)*

LAPX 260
AAR204
▲ Linde acquired this car from General American in April 1943. There were 50 of these cars built with Duryea cushion underframes and numbered as LAPX 220-269. Here is an example found in September 1969. Both this car and the one below have 3,376 cu.ft. markings, although the insides of all of these "XT" box cars were devoted to internal tankage.
(Rail Data Service, Jim Kinkaid collection)

LAPX 371
AAR204
▶ Another General American-built car. It was manufactured in 1944 and was supplied in August of that year as one of 58 from the series 326-383. Rebuilt in April 1963 it sits at Huntsville, Alabama circa 1966.
(Bernie Wooller)

LAPX 2023, 2038 & 2047
AAR204

▶▼▼ Linde was the largest producer in the world of industrial gasses and was one of the original companies that merged to form Union Carbide. The upper two photos were taken in October 1978 while the cars were sitting on the Louisville and Nashville railroad at Huntsville, Alabama. And in March 1979 the lower car was also found there. All of these cars were constructed by General American in February 1952 under Build Order 8030. They were part of the series 2001-2065 and have 4,405 cu.ft. markings since the cars were both wider and taller inside than the cars opposite. *(all, Bernie Wooller)*

LAPX 3003 & 3035
AAR204

▼▼ Two cars from the LAPX group 3001-3199. They were furnished by General American in April through July 1958 under Build Order 8123. LAPX 3003 was at Huntsville, Alabama in March 1974. At middle is a bonus: besides yet another Linde paint scheme, we can see the argon tank installation through the doorway opening. This illustration was taken in Baltimore, Maryland in February 1986.

(top, Bernie Wooller; middle, Jim Rogers)

RVCX 4316

◀ An ex-General American express car at Manitowoc, Wisconsin in May 1976. It was owned by the Richter Vinegar Corporation, the corporate successor to the A.M. Richter & Son's company of Manitowoc. After years of using wooden tanks, they acquired four of these cars. This car, along with 3316, was first mentioned in the April 1970 equipment register. They joined 1316 and 2316 already on the roster. There were two stainless steel tanks inside that carried 8,000 gallons.

(Jim Rogers)

SERX 960, 963 & 973
AAR204
▲▶▼ Shippers Car Line leased cars to the Linde company under these reporting marks. The three cars on this page are from the 930-993 set. As such they had been assembled in November 1947 under builder's lot 3229. Car 960 was found in Huntsville, Alabama in March 1979. The middle illustration was taken at River Rouge, Michigan in October 1964. The third car was at Tullahoma, Tennessee in April 1968.
(top, Bernie Wooller; middle, Emery Gulash, Morning Sun Books collection; bottom, Lee Yoder)

SERX 1028
AAR204

◀ American Car and Foundry also supplied this Linde cars. It was constructed under lot 3229A in May 1949. A part of the 994-1043 group it was found sitting in Huntsville in October 1978.
(Bernie Wooller)

AAR "TVI"

These are the "exotics" used to carry liquefied gasses. They do not have domes or manways as internal access is by access panels. Generally TVI cars carry gauging equipment to monitor and load/unload them. While most cars of this design were built with stub sills, here are two that exhibit full underframe center sills.

LTCX 5502
AAR204W

▲ While most cars of this design were built with stub sills, in this photo a full underframe is evident. The car, part of the group 5501-5506, was passing through Tullahoma, Tennessee in February 1968. A 9,000 gallon car in liquid oxygen service, it was assembled at ACF's Milton, Pennsylvania plant in November 1966. *(Fred Kite, Lee Yoder collection)*

UCOX 154
DOT113B120W

▶ Union Carbide owned this car. Found in Reading, Pennsylvania on April 22, 1987, it was constructed in March 1966 as part of the set 150-184. These 35 cars were of 30,000 gallon capacity. Note that at the time it was first listed in the register, the car was reported as class TPI, even though it is clearly marked as having a vacuum jacket and carries liquefied ethylene.
(Craig Bossler)